# International Maritime Transport

T0252649

The importance of the international maritime transport industry is difficult to overstate. This new book presents an interdisciplinary approach from a wide range of internationally based experts.

*International Maritime Transport* includes chapters on topics across the spectrum of integrated subjects (non-sector specific subjects) supplemented with contemporary articles and papers. Topics covered include: ports as interfaces; logistics; manpower and skills; financial risk and opportunities; the regulatory framework. Each part contains an introduction which explains the context of the chapter within the book and the contemporary state of the art.

Under the robust editorship of maritime experts Heather Leggate, James McConville and Alfonso Morvillo, the book is sure to be of great interest to students and academics working on maritime studies, as well as being extremely useful to professionals and policymakers in the maritime industry.

**Heather Leggate** is Director of the Centre for International Transport Management at London Metropolitan University. She is Fellow of the Institute of Chartered Accountants in England and Wales, Specialist Adviser to the House of Commons Transport Select Committee and Executive Director of Sea and Water. She has published widely in the field of maritime economics and finance. **James McConville** is Professor at the Centre for International Transport Management, London Metropolitan University, UK. He is the author of the definitive *The Economics of Maritime Transport: Theory and Practice*, and editor of the journal *Maritime Policy & Management*, which is also published by Taylor & Francis. **Alfonso Morvillo** is Director of the Institute for Service Industry Research in Naples. He is the co-ordinator of several international projects in the field of logistics and lectures in shipping economics at the Istituto Universitario Navale in Naples.

**Routledge advances in maritime studies**
*Edited by H.D. Smith*

# International Maritime Transport
Perspectives

**Edited by Heather Leggate,
James McConville and Alfonso Morvillo**

Routledge
Taylor & Francis Group

LONDON AND NEW YORK

First published 2005
by Routledge
2 Park Square, Milton Park, Abingdon, Oxon, OX14 4RN

Simultaneously published in the USA and Canada
by Routledge
711 Third Avenue, New York, NY 10017

*Routledge is an imprint of the Taylor & Francis Group*

Transferred to Digital Printing 2009

© 2005 Heather Leggate, James McConville and Alfonso Morvillo
for selection and editorial matter; individual contributors their
contribution

Typeset in Times by Wearset Ltd, Boldon, Tyne and Wear

*British Library Cataloguing in Publication Data*
A catalogue record for this book is available from the British Library

*Library of Congress Cataloging in Publication Data*
A catalog record for this book has been requested

First issued in paperback 2012

**Publisher's Note**
The publisher has gone to great lengths to ensure the quality of this reprint
but points out that some imperfections in the original may be apparent.

ISBN13: 978-0-415-64925-4 (PBK)

ISBN13: 978-0-415-34990-1 (HBK)

# Contents

# Figures

# Tables

# Acknowledgements

The publishers would like to thank Taylor and Francis Journals (www.tandf.co.uk/journals) for their kind permission to reprint:

'Editorial: Financial Creativity' by McConville, *Maritime Policy & Management*, 28, 1, 1–2 (2001);
'Investment Strategies in Market Uncertainty' by Bendall and Stent, *Maritime Policy & Management*, 30, 4, 293–303 (2003);
'International comparison of market risks across shipping related industries' by Kavussanos *et al.*, *Maritime Policy & Management*, 30, 2, 107–22 (2003);
'Perceptions of foreign exchange rate risk in the shipping industry' by Akatsuka and Leggate, *Maritime Policy & Management*, 28, 2, 235–49 (2001);
'The future shortage of seafarers' by Leggate, *Maritime Policy & Management*, 31, 1, 3–13 (2004);
'Finding a balance' by Thomas *et al.*, *Maritime Policy & Management*, 30, 1, 59–76 (2003);
'Maritime legislation' by Li and Wonham, *Maritime Policy & Management*, 28, 3, 225–34 (2001);
'Cooperation and competition in international container transport' by Meersman and Van De Voorde, *Maritime Policy & Management*, 28, 3, 293–305 (2001);
'Raising world maritime standards' by William O'Neil, *Maritime Policy & Management*, 31, 1, 83–6 (2004).

Every effort has been made to contact copyright holders for their permission to reprint material in this book. The publishers would be grateful to hear from any copyright holder who is not here acknowledged and will undertake to rectify any errors or omissions in future editions of this book.

This book is the product of a long and successful collaboration between colleagues from London, Naples, Antwerp, and Paris. However, it also draws extensively on the intellectual labours of the worldwide maritime community, who are specifically acknowledged in the text of this volume. It gives us great pleasure to record our gratitude and thanks to all those who have contributed.

*Heather Leggate*
*James McConville*
*Alfonso Morvillo*

# Biographies of editors

## Book editors

### Heather Leggate

Dr Heather Leggate is Director of the Centre for International Transport Management at London Metropolitan University. She came into academia following a professional career with PricewaterhouseCoopers and IBM. She is Fellow of the Institute of Chartered Accountants in England and Wales and a Freeman of the City of London. Her research interests are wide-ranging, notably capital market investment, risk management, UK ports policy and labour issues. She is involved with a number of institutions including the International Labour Organisation (ILO), UK House of Commons as Specialist Adviser to the Transport Select Committee and Sea and Water as an executive director.

### James McConville

James McConville is London School of Foreign Trade Professor and Visiting Professor at a number of leading institutions worldwide. His distinguished academic career, following an early working life at sea, includes pioneering research on seafaring issues and maritime economics. He is currently Chairman of Sea and Water, promoting water freight transport in the UK and is Specialist Adviser to the House of Commons Select Committee for Transport. His extensive list of publications comprises the definitive *The Economics of Maritime Transport: Theory and Practice* and he is also editor of the leading academic maritime journal *Maritime Policy & Management*.

### Alfonso Morvillo

Alfonso Morvillo is Director of the Institute for Service Industry Research (IRAT) of the National Research Council (CNR) in Naples. He is scientific co-ordinator of several international and national research projects in the field of logistics and freight transport, funded both nationally and internationally. His extensive research in the fields of logistics, supply chain management, shipping and port strategy have resulted in a large number of academic publications. He has lectured in shipping economics, management and financial analysis at the Istituto Universitario Navale of Naples. As a board member he has been involved with scientific conferences around the world.

## Perspective editors

### Patrick Alderton

Following a sea career as a Master Mariner, Patrick Alderton joined the academic team at what is now London Metropolitan University in the 1970s. In the late 1980s he was invited to take up a chair in Ports and Shipping at the World Maritime University in Malmo. On his return to London he became Visiting Professor at a number of renowned institutions. He has published a large number of articles, books and papers including the standard introductory textbook *Sea Transport – Operation & Economics*, now in its 5th edition and, more recently, *Port Management & Economics*.

### Valentina Carbone

Valentina Carbone is a Research Fellow at DEST (Département d'Economie et Sociologié des Transports), INRETS (Institut National de Recherche sur les Transports et leur Sécurité). She is a lecturer in Supply Chain Management on the International MBA, ESCP (Ecole Supérieure de Commerce de Paris) and on the Masters in 'Transports Internationaux', Université Paris I-Sorbonne. She has published widely in the area of logistics and is a representative of the Association of European Transport.

### Pietro Evangelista

Pietro Evangelista is researcher in logistics and transportation at the Institute for Service Industry Research (IRAT) of the Italian National Research Council (CNR). His research activity focuses on strategies of shipping lines and the impact of information and communication technology in his chosen field. This work is reflected in the number of papers and articles published in books and national and international journals. He is a lecturer in business economics and management at the Faculty of Engineering of the Naples' University Federico I and serves on the editorial board of two international transport and logistics journals.

### Hilde Meersman

Hilde Meersman is a professor in economic modelling at the University of Antwerp. In additions, she teaches at the Institute for Transport and Maritime Management Antwerp (ITMMA) and at the Technical University of Delft. Her involvement in international maritime transport has led her to take up the chair of the International Scientific Committee of the World Conference on Transportation Research Society (WCTRS). She is associated with a large number of research projects in areas such as international investment, modelling and forecasting freight transport, empirical analysis of port competition, inland navigation, mode choice, and sustainable mobility.

### Eddy Van De Voorde

A professor at the University of Antwerp, Eddy Van De Voorde is also visiting professor at several other distinguished institutions worldwide. His specialist fields are maritime, port and air transport economics on which he has published widely and recently co-authored a major textbook *Transport Economics*. He directs for a number of international research projects and sits on the editorial boards of *Maritime Policy & Management* and *Transport Research*. Other roles include Chairman of the Benelux Interuniversity Group of Transport Economists and Vice Chairman of the executive board of ITMMA.

# Introduction

*Heather Leggate, James McConville and*
*Alfonso Morvillo*

This work surveys recent and potential developments in maritime transport. In so doing it considers issues created by the dynamic that is ever
present in the shipping and port industry and their wider institutions. The
industry has experienced, and will presumably continue to experience,
substantial change as a result of shifts in demand for seaborne trade and
changes in the pattern of that trade. Such changes are heightened by
national participation and technological responses to economic, political
and other pressures. It is against this complex background that the present
volume is written and compiled.

It was imperative that the editors reflected the international nature of
the industry. Our aim was to create an arena within which the various perspectives on the industry could be demonstrated. Each of these contributes to an understanding of a part of the maritime industry, but none
can adequately explain the whole let alone encompass the wider environment in which it operates.

The examination of these different perspectives constitutes a multidisciplinary approach which has considerable advantages. It could be argued
however that a focus on a single discipline offers greater rigour obtained
by using the structure and techniques of an established discipline which of
necessity operates within well-defined limits. However, it is a truism to say
that all disciplines are value based and it is impossible for them to function
free from some form of bias. Any specific discipline for example, maritime
economics is value laden and it would be obtuse to pretend that it cannot
be approached as if bias free. With this firmly in mind, the present editors
agree with the general view by Allan Flanders,[1] that 'theoretically speaking these disciplines tear the subject apart by concentrating attention on
some aspect to the exclusion or comparative neglect of others'.

Approaching the subject using different perspectives utilises tools
drawn from a diversity of disciplines. This too has its specific problems.
Academics and practitioners must be conscious that a multidisciplinary
approach involves a struggle with the diverse values that each perspective
brings to the collective pool. In this lies the challenge and significant gains
from the integration of theory and practice within the industry. The

prospect of such potential gains lead the editors to embark on this venture of collating this selection of works that span a range of views and perspectives on international maritime transport. Efforts to portray in some generalised way the characteristics of the maritime industry founder on the extraordinary diversity and number of interest groups operating within it.

Superimposed on this is the instability and cyclical structure of the freight rate markets. The expanding international seaborne trade is increasingly differentiated in terms of economic, and basic physical characteristics of the commodities shipped, the size of the consignment as well as the pattern of trade. Such trends affect different sectors of the industry in different, often selective, ways. For these reasons the industry is, in reality, not a single entity but what can be termed the cluster of related industries differing in the demand they satisfy and the technology they use. With the inclusion of ports this differentiation is further heightened. In producing this book, the editors took cognisance of the challenge that such diversity presents and have tried to make it as comprehensive as possible within the physical limitations of the volume.

Conscious of this problem, the choice of consistent articles was not made easier by the recent plethora of high quality work on the subject. The central issue was one of presenting the traditional mainstream perspectives of the industry while recognising the need to embrace the issues, nuances and insights. Much of this book comprises original contributions from maritime specialists around the world, with a limited number being drawn from recent issues of *Maritime Policy & Management*. This journal, a unique publishing venture now in its fourth decade, has been at the centre of the literary explosion around international maritime transport.

This volume represents a radical departure from previous works in its structure and approach. Each section deals with a different perspective on the maritime industry and is edited by experts in that particular field. The section editors each discuss the state of the art in the opening chapter before introducing a selection of works providing a wide-ranging analysis of the subject. Wide discretion of approach has provided literary freedom for individual opinion and analysis within the overall framework. This permits a level of innovation which is perhaps stifled in a more standardised model. While each section or perspective can be seen as exclusive, together they all form a comprehensive volume of issues in contemporary maritime transport. To this end, the book has also been subject to overall scrutiny to ensure cohesion and consistency.

The first perspective, edited by Heather Leggate, endeavours to range over the field of financial strategy. As is well known, the problems of defining finance, and for that matter capital, are legion and to achieve unanimity as to its scope and content is virtually impossible. With this in mind, the author wisely concentrates on examining finance in a business operational context, as a function underpinning the day-to-day operational or investment activities. The inbuilt traditional attitudes and aspirations of the maritime industry are questioned in terms of finding appropriate forms of finance, accessing the capital markets and risk management opportunities. It thus highlights the need for entrepreneurial skills when financing successful ventures in the uncertain and volatile climate in which this industry operates.

Labour is in the broadest sense the active and principal factor of production, without which nothing can be accomplished. Combined with the other factor of capital it produces maritime services and its contribution to this process is analysed in the 'People and skills' perspective edited by Valentina Carbone. The emphasis throughout is on the fact that it is a differentiated resource and needs to be allocated in such a way as to ensure the most efficient use of the available and potential supply. It is a supply influenced by an extremely wide variety of conditions and one that cannot be simplified either at a theoretical or practical level as a homogeneous factor. The central question emerging from the present discussion is the unique mobility of seafarers. Occupational mobility is controlled by institutions external to the industry. Certificates of competency, for example, are issued by national governments conforming to international regulation. Once such qualifications have been secured they facilitate mobility within the industry but have only a minimal relevance to alternative non-maritime occupations. Seafaring necessitates geographical movement and to some extent migration. But while non-national seafarers participate in other countries' labour markets they achieve this usually without entering the country of the flag registry by which they are employed. This has resulted in the labour market for seafarers becoming increasingly fragmented on a geographical basis.

The discussion of the port industry highlights the vital contribution it makes to the national and international economy, levels of commercial activity and general social wellbeing. The industry has experienced and continues to experience deep-rooted changes engendered by the economic dynamics of international seaborne trade. Considerable knowledge of these changes, which have been technological, institutional and regulatory in character are analysed in the 'Ports in transition' perspective edited by H. Meersman, E. Van De Voorde and T. Vanelslander. What is made absolutely explicit is the symbiotic relationship between ports and shipping, a relationship emphasised by the extraordinary changes occurring in recent years. These changes emanate almost entirely from the shipping sector. The prime example of this is the impact of the container, which has been almost totally confined to liners and ferry operators, thus limiting much of their impact to port structures, often wholly committed to the particular requirements of container traffic. Port owners and operators have up to now largely reacted to exogenous factors in the development of their strategies. The authors suggest that ports are increasingly adopting a more proactive approach in order to be commercially successful and efficient.

The maritime industry has in recent times significantly increased its specialised functional base. At the centre of this has been unitisation and containerisation. Originally envisaged as a comparatively simple systems innovation, it has grown in complexity, the repercussions are highly significant. Essentially, this has entailed a movement away from concentration on a particular transport mode to one preoccupied with the quality of services and specifically the time element in the whole transport process. The necessary technological and managerial change has created what is known as intermodalism out of which grew physical distribution management, which spawned logistics and, in a wider context, supply chain management. The basic objective of logistics operations is to secure and continue to achieve economies of scale mainly through the consolidation of product flows across different enterprises. The competitive pressures create efficiencies throughout the chain particularly in relation to freight handling. This is the basis of the integrated transport system whose contemporary situation and potential is examined in the perspective on 'Logistics and ICT' edited by Pietro Evangelista.

The complexity of maritime activity necessitates some form of social control. The extent of such regulation is a persistent controversy within the contemporary shipping arena. The structure, extent, and development of the regulatory framework are the concern of the final perspective 'The regulatory framework' edited by Patrick Alderton and Heather Leggate. This section concentrates on long-term controls, promulgated by national, regional and international institution of one form or another. What is made explicit is the fundamental impact of regulation on the industry's economic and social wellbeing and a need for an extensive study of the whole area. Maritime activity has an exceedingly long history of regulatory involvement, particularly in regard to ownership and safety. The latter has assumed a predominant role in recent times, resulting from periodic environmental crises created by serious maritime catastrophes, such as the loss of the *Exxon Valdez* and the *Prestige*. The discussion elaborates on the function of regulation and regulatory bodies within wider maritime activity and their substantial international role. Emphasis is placed on the necessity of an increase in the effectiveness of regulation, and specifically the need for enforcement.

This undertaking to analyse and combine a number of perspectives on one industry perhaps exposes itself to a lack of inner cohesion and repetition. Neither have or could be completely avoided. Indeed we would argue that this is desirable for it reflects the interrelationships between the perspectives from which coalescence follows. Some readers will not of course accept particular hypotheses or conclusions proffered. Our aspiration is that this will spur them to contribute to the development of a multidisciplinary approach to the maritime industry. This in our view is the only way it can be examined holistically. Whether the proper balance has been struck between technical accuracy, theory and its application is for the reader to judge. We would concur with Dr Johnson's famous comment that no work of such multiplicity is ever free of 'a few wild blunders, and risible absurdities'.[2] We are therefore impelled in the time-honoured fashion to take exclusive responsibility for the shortcomings and imperfections encountered in the following pages.

## References and notes

1 Flanders, A., 1965, *Industrial Relations: What is Wrong with the System?* (London: Faber and Faber), 9–10.
2 Preface to *The English Dictionary*, 1755, paras 51–94.

# Part I
# Financial strategy
*Edited by Heather Leggate*

# 1 Financial risks and opportunities

*Heather Leggate*

## 1 Introduction

The Shipping Industry operates in an environment of uncertainty, with the cyclical and unpredictable nature of freight rates and the changing fortunes of the second-hand market creating a background of instability against which the business must operate.

This level of risk has implications both for the corporate entity and its investors. From the corporate perspective, the level of business risk has to be taken into consideration in appraising future investment, and obtaining appropriate sources of finance for those investments. As far as the investor is concerned, a high level of risk is expected to translate into high return.

Further financial risks are generated by the level of debt finance, whilst the volatility in the exchange rates particularly against the US dollar and fluctuations of interest rates can have a significant impact on performance.

But what of the opportunities? Shipping is an industry steeped in tradition and often it is these traditional attitudes which prevent a more imaginative approach to financial management.

Shipping is generally recognised as being a low margin business. Income is derived from the movement of cargo, which is dictated by the freight rate. Expenses comprise voyage costs, port dues, administrative costs and wages which may be substantial, leading to low operating profit. Given the huge capital requirement in terms of vessels, financing costs in the form of interest can further reduce the overall profitability of the business. Typically, the industry has used substantial amounts of debt to finance its activities and therefore has what is known as high gearing or leverage. This huge financing requirement is a reflection not only of the capital intensity of the industry but also the ageing fleet.

## 2 Sources of finance

Sources of finance available to companies essentially fall into two categories: debt and equity. Both debt and equity have a cost. Companies strive to minimise this cost in the establishment of their capital structure. Debt theoretically has a lower cost than equity because it carries less risk to the investor. If this is the case, why do companies bother with equity finance? The answer is that the financing decision is based on a number of factors not merely the costs of capital but control of the business, cash flow and access to funds. It could be argued that financing is principally a marketing problem. The company tries to split cash flows generated by its assets into different streams that will appeal to investors with different tastes, wealth and tax profiles.

### 2.1 Equity

The equity holders or shareholders are the owners of that business and have ultimate control of the company's affairs. In practice, this control is limited to a right to vote on appointments to the board of directors and a number of other matters. They hold the equity interest or residual claim, since they receive whatever assets or earnings are left over in the business after all its debts are paid. However, they have limited liability in that the maximum amount they can lose if the company goes bust is their investment in the shares. None of the other assets of the shareholder is exposed to the company.

The cost of equity to the company is the return derived by the shareholders. This takes the form of dividends and capital gain. Dividends are usually paid annually or semi annually out of earnings or profits and are discretionary. Indeed, a profitable business may decide to retain earnings to reinvest in the business. These retained earnings are in fact a major source of finance for a company. Capital gains or losses may arise from fluctuations in the share price and may be realised if the holder sells his shares. Since both dividends and capital gains are uncertain, shareholders require a high rate of return to compensate them for this risk or uncertainty which can make equity a costly form of finance.

The issue of share capital is also more costly than the issue of debt due mainly to the exacting compliance procedure relating to stock markets, particularly on the established exchanges. There are however, cheaper ways of doing this through the issue of shares to existing shareholders or by the placement of shares to known investors.

Apart from the relatively high cost of equity, another reason why shipping companies have avoided equity finance is that it may mean relinquishing control of the business. This is a particular problem for the smaller companies where the shares are owned by family members. An increased number of shares distributed to new investors will dilute the overall voting power of existing shareholders. One way of avoiding this situation is to make what is known as a 'rights issue'. Here the shareholders are offered additional shares often at a discount on the market price in proportion to their existing holding. If the 'rights' are taken up, the percentage holding remains unchanged, but the company still manages to raise the required finance.

## 2.2 Debt

Lenders are not the owners or proprietors of the business and therefore have no voting power. The variety of corporate debt instruments is almost endless. The instruments are classified by maturity, repayment provisions, seniority, security, default risk, interest rates (fixed or floating), and issue procedures. From the investment perspective, debt carries less risk than equity because interest and principal must be paid before there can be any distribution to the shareholders. Debtholders are entitled to a fixed regular payment of interest and repayment of the principal according to an agreed schedule. Interest may be fixed over the period of the loan regardless of what happens to the prevailing rate of interest or may be variable rates often tied to the LIBOR (London Interbank Offer Rate) which is the interest rate at which major international banks in London lend dollars to each other. If the company fails to make these payments, it defaults on the debt and it can file for bankruptcy. The usual result is that debtholders then take over and either sell off the company's assets or continue to operate under new management. Default is therefore the most important risk for lenders or investors in debt instruments.

The risk of default is quantified by rating agencies notably Standard and Poors and Moodys. These ratings are based not only on historical information but an impression of the future. Both these rating agencies see the industry as high risk because of its economic sensitivity, capital intensity and competitive structure. The key factors in the analysis are capital structure, age, size and diversity of the fleet, management ability. Capital structure relates to financial gearing, or the proportion of long-term debt finance. The industry has traditionally relied on loan finance and as such has high levels of gearing, often in excess of 70 per cent. Clearly this factor contributes to the possibility of default in periods of recession. The age of the fleet has already been discussed in terms of capital requirements, but it is also an issue for the investor since, without detailed knowledge, it is an indicator of quality. The running costs are also higher for older vessels, which has a cash flow implication. The size of the fleet affects the rating. Companies with larger fleets tend to achieve a higher grade than those with smaller fleets. Associations with other companies, for example, a highly rated parent company, or a highly rated charterer also affect the decision. The diversity of the fleet in terms of operating sector is a further issue for the reduction of market risk. Finally, sound management and industry expertise is examined in considerable detail to determine the future prospects for the company. These are the factors that affect perception of the company for both the rating agencies and the investor, and ultimately determine the cost of the finance to the company. Historically, the grades achieved by the shipping issues have been poor.

The cost of debt is determined by the interest rate or coupon on the debt instrument. As previously explained, debt should be cheaper than equity because of the certainty of return. From a corporate perspective, the cost is further reduced by taxation, since tax authorities treat interest payment as a cost. That means that the company can deduct interest when calculating its taxable income. Thus, interest is paid from pre tax income and dividends from post tax income.

Currently interest rates are at 40-year lows and lenders are queuing up to throw money at the shipping industry. Further, the consolidation among the large industry players such as Wallenius Wilhelmsen and Hyundai Merchant Marine car carriers, Bergesen, Teekay Shipping and Navion has created the need for larger and larger levels of debt.

### 2.3 Bank finance

Bank finance remains the most important method of raising capital in the shipping industry. Financial institutions provide loans in varying forms to shipping companies. At the simplest level, they are term loans under which the bank lends a certain amount to the shipping company to acquire a vessel over a specific period. The loan is repaid over the period using the income generated by the vessel and sometimes its residual value. These repayments are made up of the loan principal and interest. Such loans can be customised to the requirements of the user. This can include equal or unequal instalments, and even periods of moratoria whereby the capital element of the loan is suspended. Such flexibility can be helpful in deteriorating market conditions since the shipping industry is very cyclical leading to volatility in earnings. Interest is normally paid at a variable rate based on LIBOR (London Interbank Offer Rate). LIBOR is a base line rate and the company will typically pay a percentage in excess of this rate. Fixed rate loans are less common.

In order to minimise their risk, banks may ask for collateral, which may take various forms. Such security may be direct in the form of a mortgage on the vessel itself or indirect in the form of guarantees. Syndicated loans are also becoming more widespread as a method of diversifying risk. In this situation, a number of financial institutions will share the loan and thereby share the accompanying risk. The number of participating banks can be very large but in the shipping industry it is generally between four and eight banks all having shipping expertise and experience.[1]

The cost of a bank loan is the interest payable on the debt, which will depend on the particular deal struck with the financial institution.

*2.4 Bonds*

Bonds represent an alternative form of debt finance, which are increasingly used by the industry. The debt is essentially carved up in small bundles and sold to a number of investors on the capital markets in a similar way to shares. The fixed amount of interest paid to the lender is known as the coupon and the loan is said to 'mature' on a specified date.

The price of a bond fluctuates over its life and is inversely related to the prevailing rates of interest. At any stage an investor can sell his or her bond at the prevailing market price and in so doing make a capital gain or loss.

Bonds should have a number of advantages over both traditional bank finance and equity. Firstly, in comparison with a bank loan, there is a positive cash flow effect because the principal is not amortised. In other words, the company only has to pay the 'interest' element of the loan during the life of the bond. Repayment of the principal is held over until the end. The coupons themselves are typically paid on a semi annual basis, which again can ease cash flow. Furthermore, the company has the option to rollover the debt at the end of the life of the bond, for example issue more bonds or shares to cover the repayment. Another possibility would be the use of convertible bonds, which, if converted, become permanent equity capital, which does not have to be repaid at all. In comparison with equity, debt finance should cost less, as it is a less risky form of finance. Coupon payments are fixed and there is a guaranteed maturity value. Furthermore, the tax deductibility of debt makes it even more cost effective.

The industry's initial flirtation with the bond market in the late 1990s proved difficult with investors demanding high coupons reflecting low credit ratings. Since then there has been a marked improvement with the industry raising $1.3 billion worth of new issues since 2001. The total for 2003 alone was more than $550 million including the CMA-CGM issue of the first shipping junk bond denominated in Euros. The position in the bond market stems from the high liquidity, that is the large amount of trading that is taking place.

## 3 Capital structure

In practice, companies use a mixture of debt and equity finance. Gearing or leverage is an indicator of that structure. More precisely, gearing is a measure of the extent of debt capital used to finance the business. The existence of debt increases the financial risk, which must be borne by the shareholders. This risk stems from the fact that interest on debt must be paid before the shareholders can receive a dividend. Failure to meet this fixed commitment can mean bankruptcy for the company. A highly geared company will also find it difficult to obtain further debt finance because of the risk of bankruptcy. To compensate them for this risk they require a higher rate of return.

Wikborg (1993)[2] estimates gearing for shipping companies at around 65 per cent on average with a typical variation of between 50 per cent and 80 per cent depending on trade, company and country. The optimal capital structure for a shipping company depends on a number of factors such as tax exposure, stability of trade, length of charter, and the shareholder risk profile. Bankruptcy is also a key issue given the number of companies experiencing financial distress over the last 30 years. He suggests the ideal gearing would be around 50 per cent, but that in reality this is neither achievable nor sustainable. One of the main reasons for this is the apparent lack of knowledge and transparency within the industry, the fact that 'being an insider is perceived by many as necessary to exploit the business cycles and risks'. It is therefore the nature of the industry, which prevents adequate levels of equity investment, and hence high gearing is a requirement for the financing of operations. Leggate and Carbone (2001)[3] in a study of 19 quoted companies worldwide found that gearing is not as high as many believe at around 43 per cent. However, there was a clear lack of uniformity across the industry, with results varying greatly from company to company and (with the exception of Japan) across country.

Theories of capital structure suggest a strong relationship between gearing and the return on equity, as shareholders require compensation for the financial risk suffered. The first of these theories is known as the traditional theory, which argues that the relationship is non-linear. In other words, shareholders require increasing returns to financial risk.

The assumptions, which underlie the model, are:

- perfect capital markets;
- rational investors;
- homogeneous expectations; and
- a corporate goal of maximising shareholder wealth.

For low levels of gearing, shareholders do not expect a high degree of compensation for financial risk. As gearing increases however, their required rate of return rises. The cost of debt is constant regardless of the level of gearing. The Weighted Average Cost of Capital (WACC)[4] takes account of the costs of equity and debt weighted by market values. This is initially reduced by the low cost of debt, but as the debt finance increases, the higher return on share capital outweighs the cheapness of the debt, and so the WACC starts to increase. The minimum WACC indicates the optimal capital structure.

The traditional theory (Figure 1.1) was brought into question by Modigliani and Miller (1958)[5] who presented a theory whereby the capital structure was irrelevant to the overall cost of capital.

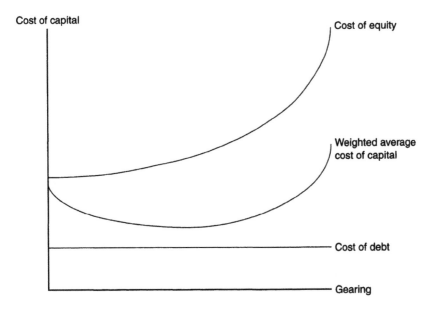

*Figure 1.1* The traditional theory.

The additional assumptions of their model are:

- individuals and firms can borrow at the same rate of interest;
- all bonds have identical yields; and
- interest rate does not vary with the degree of indebtedness.

Modigliani and Miller made the propositions that the value of companies with the same amount of business risk will be equal regardless of the capital structure, and the cost of equity or return on equity will increase as a linear function of the gearing ratio. The reason for this is because of a process of arbitrage, whereby investors can increase their income by taking on personal gearing equal to the gearing levels of the geared company. Thus, if two similar companies have different market values then the investor would move out of one into the other until the values were equalised. The model was later adapted to include the effects of taxation, which makes debt more attractive because of the tax deductibility of interest payments.[6]

Figure 1.2 shows the same constant cost of debt, but the cost of equity increases in a linear fashion such that the WACC remains constant regardless of the level of gearing.

Both models suggest a relationship between cost of equity and gearing.

Shipping is a highly capital intensive industry with a huge financing

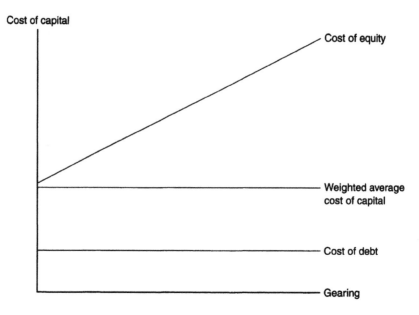

*Figure 1.2* The Modigliani and Miller model (without taxation).

requirement. It is also an industry which is subject to rapidly changing fortunes. From the perspective of an investor or financier, this volatility of earnings creates the risk that the costs of finance will not be met. This risk is somewhat alleviated by the fact that the assets are of very high value and provide security to the providers of capital. Shipping companies have traditionally looked to bank loans due to their relatively low cost. However, they are increasingly using more progressive sources of finance, such as equity and bonds, in order to access higher levels of funding with cash flow benefits. As consolidation in the industry continues we can expect to see greater access to the capital markets by shipping business.

## 4 Selected works

The works in this section deal with finance from both the corporate and investment perspectives. 'Financial creativity' formed an editorial by James McConville[7] in which he observes that maritime finance receives little academic attention despite the capital intensity of the industry. Greater research in this area may encourage a more imaginative approach to financing activity. The corporate theme continues in 'Investment strategies in market uncertainty' by Bendall and Stent.[8] Here they discuss the shortcomings of discounted cash flow techniques for project appraisal because it takes no account of changing cash flows in dynamic market conditions. They suggest that Real Options Analysis (ROA) incorporates such contingencies and that the flexibility itself has an intrinsic value which must form part of the decision process.

The final two chapters of this section consider perception of the industry from an investor standpoint. Kavussanos, Juell-Skielse and Forrest[9] analyse the risk inherent in the shipping sector as indicated by returns on shares. More specifically they measure the systematic or market risk ($\beta$) which cannot be 'diversified away' in the investment portfolio. Surprisingly, they find that shares in the Bulk, Tanker, Container and Ferry sectors have lower than market risk whilst that of the Cruise industry carries the same risk as the market as measured by the Morgan Stanley All World Index. The results however are based on the relatively small numbers of quoted shares and as the authors suggest there is a need for further study when a larger sample is available. In the final chapter, 'Perceptions of foreign exchange rate risk in the shipping industry', Akatsuka and Leggate[10] focus on exposure of the industry to exchange rate fluctuations. Any international business suffers this risk, but the Shipping Industry is particularly susceptible in that its freight income is predominantly US dollar based, whereas costs must be met in a variety of other denominations. Based on a comparison of Japan and Norway. The article highlights the differing level of exposure in the two industries and the impact of exchange rate fluctuations of the yen and the krone against the US dollar on the share prices of the two industries. It finds that the impact is greater for the Norwegian industry, which maintains a higher level of exposure and less for the Japanese. It is also interesting to note the dramatic affect of the Asian Crisis on the Japanese company results.

All these chapters highlight the fact that financial issues are central to any business strategy. The Shipping Industry is clearly aware of the risks, but has an opportunity to show a greater scope and creativity in their management.

## References and notes

1 Grammenos, C.Th., 2001, Credit risk, analysis and policy in bank shipping finance, C.Th. Grammenos (ed.) *Handbook of Maritime Economics and Business*, (London: LLP).
2 Wikborg, C., 1993, Financial gearing in shipping – is it too high? Can it be reduced?, selected papers from the 5th International LSE Shipping Conference 1992.
3 Leggate, H.K. and Carbone, V., 2001, Financial characteristics of the shipping industry: challenging the preconceptions, IAME Conference Proceedings, Hong Kong, July.
4 The WACC is weighted by market value of debt and equity to reflect the most current position.
5 Modigliani, F. and Miller, M.H., 1958, The cost of capital, corporation finance and the theory of investments, *American Economic Review*, 48, 261–97.
6 Modigliani and Miller modified their model to take account of the fact that tax

relief is available on interest payments. This makes debt even cheaper and means that the value of the geared company will be higher than that of a similar ungeared company by the amount of the tax shield. (The tax shield is the tax multiplied by the interest on debt discounted to the present day with the same interest rate, i.e. $\dfrac{t \times i \times D}{i}$)

The propositions there become:

- $Vg = Vu + Dt$

- $Keg = Keu + \dfrac{D}{E}(1 - t)\{Keu - Kd\}$

- $WACCg = WACCu \left(1 - \dfrac{Dt}{D + E}\right)$

7 McConville, J., 2001, Editorial: financial creativity, *Maritime Policy & Management*, 28, 1, 1–2.
8 Bendall, H. and Stent, A., 2003, Investment strategies in market uncertainty, *Maritime Policy & Management*, 30, 4, 293–303.
9 Kavussanos, M., Juell-Skielse, A. and Forrest, M., 2003, International comparison of market risks across shipping-related industries, *Maritime Policy & Management*, 30, 2, 107–22.
10 Akatsuka, K. and Leggate, H.K., 2001, Perceptions of foreign exchange risk in the shipping industry, *Maritime Policy & Management*, 28, 3, 235–49.

# 2 Financial creativity

*James McConville*

An examination of *Maritime Policy & Management* and indeed other contemporary shipping literature showed the same paucity of articles and books in this area. This apparent lack of importance attached to finance and investment is also evidenced by the fact that many maritime postgraduate programmes deem it sufficiently marginal to exclude from their syllabi.

The importance of finance may easily be illustrated by forecasts of capital requirements given ageing tonnage and international seaborne trade expansion. The average age of the world fleet is nearly 15 years, with 52.6 per cent over this threshold. On sector analysis, the age distribution is similar, being skewed to the highest age category in all cases. Of these, the general cargo ships have highest average with 64.5 per cent over 15 years old. Tankers also have a higher than average percentage over 15 years. Such estimates recently produced for the world fleet indicate a newbuilding finance requirement of US$125 billion, and second-hand of US$33 Billion. This does not take into consideration working capital and restructuring needs.

The issue of high capital demand is compounded by the tightening of credit facilities by the shipping banks following the wave of defaults in the 1980s. Poor market conditions, particularly in container and dry bulk shipping, and of course the Asian crisis have further intensified this situation with a number of banks disposing of their shipping portfolios in the latter half of 1997 and 1998, including some well established financiers. Other commercial banks have also reduced their provision as a result of poor market performance. The problem is greatest for the smaller and less established companies who have not built up a strong relationship with the banks.

There is a necessity for more imaginative forms of finance, and it would appear that the industry has not yet embraced this challenge. Thus far, the use of the capital markets is extremely limited with some 300 equity and only 30 bond issues. A major problem is the relatively small market capitalization of the industry, which means that it attracts limited attention from market analysts.

The monitoring of financial performance is even more pressing in an industry typified by low margins. Strong financial management can mean the difference between profit and loss. The traditional attitude is such that the industry does not see its role in speculation, yet it is a necessary part of modern risk management strategy.

The significance of finance is clear. Why then does it receive such limited academic attention? The answer probably lies in the fact that many players in the industry are concerned almost exclusively with the operational side of the business. In other words, the ship and its ability to carry cargo. The enormous detail and complexity surrounding this preoccupation clouds the important fact that shipping is not merely an operation but a business. Until this fact is recognized by industrialists, academia and students there will be insufficient research in the vital financial area and the industry will remain years behind other multinational business in terms of financial sophistication.

In short, there is an urgency for greater research, which examines shipping as a business, in order to strengthen financial and business creativity.

# 3 Investment strategies in market uncertainty

*Helen Bendall and Alan F. Stent*

## 1 Introduction

Maximizing a firm's value is dependent on correct investment choices, thus management needs sound and reliable tools to minimize the risk of poor investment decisions. Because of the changing nature of global markets and the highly capital intensive characteristics of the maritime industry any investment in such a dynamic environment could be described as courageous. The industry's cash flows are exposed daily to changes in international financial markets, to multicurrency sovereign risk, to pirates, to movement in oil prices, to changes in commodity prices, to other competitors, and above all, the trade cycle. Because of the long lead times of 12 months or more between placing an order with a shipbuilder and putting a ship into service, the profitability of the investment can alter dramatically. The original investment decision, therefore, should factor in freight rate and asset price volatility and should consider possible alternate uses for the asset should the economic reality vary greatly from that envisaged when making the commitment.

Shipping is a service industry whose demand is closely correlated with international trade levels and patterns. As a consequence shipping is subject to sometimes unpredictable swings in demand requiring the operator to make strategic planning decisions while navigating through the economic cycle of many countries. While boom economies generate rising freight rates and encourage investment, ship operators may also have to face falling freight demand and declining freight rates, often deteriorating to uneconomic levels for extended periods. Ship owners and operators must make strategic long-term decisions which take account of the flow-on consequences for a service industry of varying trade cycles in an international context. Indeed, the shipping industry is often a leading indicator of movements in the trade cycle as demand for shipping varies sharply as exporters anticipate changes in demand for their products.

Thus an investment in shipping can be regarded as a large-scale capital evaluation problem within the context of a great number of volatile parameters.

Management's strategic vision of the future and ultimate value of the firm can only be realized if the project has been valued correctly as ships are usually major capital investments with significant implications for the balance sheet. Investments have traditionally been valued using Discounted Cash flow Analysis (DCF) and, in particular, Net Present Value (NPV). In the investment decision process, implicit assumptions concerning an expected scenario of cash flows (CF), are made and these are discounted at the risk adjusted rate, $r$, to determine the present value of the project.

$$NPV = \sum_{n=1}^{t} \frac{cashflows_t}{(1+r)^t} - I$$

Projects are accepted, or at least considered favourably, if the discounted value of these cash flows, the project's present value (PV) is equal to, or exceeds, the capital cost of the investment (I). If the NPV value is positive the project should be accepted as it would add value to the firm. If negative, the project should be rejected as going ahead with the investment would destroy value in the firm.

The international shipping market is characterized by change, uncertainty and competitive interactions so that the implicit assumptions regarding future cash flows in the traditional approach will probably differ from what is realized. The traditional approach assumes that management, having made the decision to initiate a capital investment, will manage cash flows continuously as planned until the end of its pre-specified useful life. This approach ignores the ability of management to adapt or revise decisions in response to unexpected market developments. As new information arrives, uncertainty about future cash flows is reduced. Management may have the flexibility to alter its initial strategy to capitalize on favourable opportunities or minimize losses.

This chapter will demonstrate the use of Real Options Analysis (ROA), to enable managers to accommodate flexibility in the investment decision so that the valuation of a project can reflect operating and strategic adaptability, i.e. the traditional NPV analysis expanded to include contingencies. The principles of ROA, a literature survey and discussion of the key issues follow in section 2. Section 3 will demonstrate the efficacy of the ROA approach with an example of a typical strategic investment problem faced by a shipowner in times of economic downturn and competitive pressures. The model will be explained in section 4 followed by the results of the ROA study in section 5. The chapter will be concluded in section 6.

## 2 Principles of Real Option Analysis

Discounted cash flow (DCF) analysis for project evaluation is taught extensively in business schools and yet studies of corporate investment practices in the maritime industry[1] and elsewhere[2] have shown a divergence between traditional finance theory and corporate reality. Many argue that in practice DCF capital budgeting techniques are often implied as a set of checks rather than as the principal valuation tool[3] or simply to justify a senior managerial decision already made.[4] Often there is a lack of understanding of how to include strategic issues in the traditional analysis. That is not to say that managers do not recognize that the failure to do so may lead to costly errors, but the difficulty of incorporating such planning in a DCF analysis and a lack of understanding of how to do so lead many to ignore the potential costs and hope that serious problems do not arise.[5] Managers must be able to include uncertain future events and outcomes and potential strategic responses in a prospective analysis of a capital investment project. Traditional NPV, by failing to recognize and capture the need for flexibility to alter in some way the project in the future, can over- or under-value a project.

What is needed is an extra parameter embedded into the capital budgeting tool which would capture the flexibility to adapt and change the investment parameters in response to altered market conditions. This would expand the upside potential of an investment while limiting downside losses relative to management's initial expectations under the predetermined scenario. Such an expanded NPV analysis would incorporate both the direct cash flows from the traditional NPV and a real option value which would reflect operating and strategic adaptability.

### 2.1 Real options

The concept is similar to financial options.[6] An option gives the right but not the obligation to undertake an action at pre-determined cost (the exercise or strike price) for a pre-determined period of time (the life of the option). It gives the investor the right to defer, to expand, to contract or abandon the project once more information becomes available and uncertainty is diminished. A *call* option is the right to buy the underlying asset at the pre-determined (exercise) price. There is no obligation on the part of the holder of the option to exercise the call so the option can lapse. However, if the option is exercised then the profit on the option is the difference between the value of the underlying asset and its exercise price. A *put* option is the right to sell the underlying asset to receive the exercise price and thus is the opposite of a call. A *European* option can only be exercised on its maturity date whereas an *American* option can be exercised at any time during its life and thus is generally more applicable to investment in real assets. An option is *in the money*, i.e. profitable to exercise if the price of the underlying asset is above the exercise price with a call option and below the exercise price with a put option. If not profitable to exercise it is *out of the money*.

Before the development of ROA, managers and strategists intuitively adjusted their investment strategies to include other factors such as growth considerations, realizing that often traditional DCF criteria undervalued investment opportunities. Kester[7] developed Myres[8] concept of thinking of discretionary investment opportunities as 'growth options'. The existence of an extra factor, a growth option, Kester argued, could explain inconsistencies between capital budgeting theory and practice in the real world, particularly where there were strategic or competitive aspects involved.

In a shipping context, for example, shipping lines may enter a new market or trade not so much because the immediate investment generates a positive NPV, indeed it may be the opposite, but in order to keep a competitor out of the trade or to put the line in an advantageous position for valuable follow-up opportunities. An investment such as this is an example of a multistage decision that involves 'real' options. The decision to enter the trade has the ability to create future assets (cash flow) as a by-product of the initial investment decision. The shipping line, by undertaking the initial investment, has the option in future to expand the number of ships in the trade or exit the trade, depending on market circumstances in the future.

There is a wide body of literature in the field of real options, though 'empirical research has lagged considerably behind conceptual and theoretical contributions in the literature'.[9] Only a few papers have applied ROA to maritime applications.[10-13] Trigeorgis,[14,15] amongst others provides a good summary of the field. Many papers give taxonomy of real options by classifying them by the function they perform, i.e. expansion, contraction, abandonment, deferral, growth, strategic, competitive, switching options, etc.[11,14,15]

### 2.2 Valuing real options

Black and Scholes[16] and later Merton[17] developed the quantitative origins of real options in pricing financial options. Kasanen and Trigeorgis[18] maintain that real options can in principle be valued in a similar manner to financial options, even though they may not be traded, because the process of capital budgeting determines the value of the project's cash flows in the market. The Black–Scholes model, however, is 'complex and off-putting to many practitioners'.[19] Although mathematically challenging and some of the assumptions may be too restrictive for pricing real options, the principles are useful in developing an understanding of real option valuation. Cox, Ross and Rubinstein's[20] binomial approach presented a simplified methodology by valuing financial options in discrete time. Cox and Ross,[21] by recognizing that an option can be replicated (to create a 'synthetic' option) from an equivalent portfolio of traded securities, facilitated further the ability to value real options.[22] The replicating portfolio approach is based on the Law of One Price, which simply states that to prevent arbitrage (riskless) profits two assets with the same payoffs ('twin security'[23] or 'twin asset'[24]) in every state of nature are perfect substitutes (i.e. perfectly correlated) with the underlying risky asset and therefore have the same price (value).

Copeland and Antikarov[25] point out the difficulties of finding a market-based 'twin asset', or 'twin security', that is perfectly correlated with the underlying asset. Instead of searching for a perfectly correlated asset in the market they propose that the present value of the project itself be used as the value of the underlying asset.[26] It is this approach that will be applied in section 4. The project is equal to the static (inflexible) NPV plus a value for active management, (a real option).

Value of project with flexibility = Value of project without flexibility
+ Value of flexibility

By incorporating the value of the real option in the project evaluation a real options approach views capital investment as an on-going process requiring active management involvement.

## 3 Proposed scenario

The following scenario will provide an appropriate background example to demonstrate the efficacy of ROA in capital budgeting decisions involving strategic considerations. The example is hypothetical, although based on an actual trade with realistic data input. The scenario describes a typical investment decision problem faced by a shipowner in times of economic downturn, falling freight rates and competitive pressures.

An established shipping company 'X' has had four modern ships of 1,500 TEU capacity with service speeds of 23 knots in the Trans-Tasman trade, east coast Australia to New Zealand. The company's four ships are modern and were built in Korea, the last being delivered six months ago. The company's ships are servicing three ports in Australia and two ports in New Zealand, achieving a four-day sailing service. The trade has been growing steadily over the last few years and the shipping line has been achieving on average 90 per cent load factor. Freight rates in the Trans-Tasman trade have been variable and have fluctuated ±30 per cent over the long run, though trending downwards over a 20-year period.

Although the company has faced falling freight rates in the long term, profitability has been maintained by introducing efficiencies in operations and management. The fleet was modernized for this purpose and the company has benefited from increased vessel capacity so slot costs have fallen. The vessels with service speeds of 23 knots have also improved productivity by allowing more round voyages per year.

When the contract was signed for four ships the company was granted an option on a further vessel of the same size and speed. The option for the fifth ship will expire at the end of the month and the company is considering whether or not to exercise this option. If they go ahead with the purchase (exercise the option) they will receive delivery in 15 months time. Until recently the company had intended to do so despite a softening of demand and falling freight rates, believing that the trade was well established and that the fifth ship would allow them to increase their share of the market. However, about five weeks ago a competitor, company 'Y', entered the trade with two older 2,500 TEU ships with service speeds of 16.5 knots. These ships cannot offer the same quality service as 'X' as their round-voyage times are considerably longer than 'X's' modern fleet. 'Y' also calls at only three ports (two in Australia and one in New Zealand). These ships had been bought cheaply as they were 15 years old but the company is very highly leveraged so needs returns quickly to service its debt. 'Y' thus adopted an aggressive pricing policy.

'Y', through heavy discounting, secured 20 per cent of the trade by the second round voyage. This was achieved by targeting major shippers in Australia and New Zealand for base cargo, offering a rebate for loyalty clauses. However, 'Y' is only achieving a 50 per cent load factor and with only two ships sailing dates are less flexible.

Freight rates, in general, have been falling over the last six months due to a worsening global economic climate, even before the competitor appeared. It is not known how long this down cycle will last as economists are mixed with their projections. Cycles have not been regular in the past so the upturn will be difficult to predict. If the downturn continues too long any retaliatory pricing by 'X' and other measures to win back cargo may not be possible indefinitely for although 'X' has reserves these are finite.

The investment decision by 'X' now faced by management has a number of unknown factors. Not only does 'X' have to consider the current softening of freight rates aligned with global downturn but now is faced with an aggressive competitive environment. 'X' does not know how successful 'Y' will be and whether they will still be in service when the fifty ship is delivered in 15 months time. 'Y's' predatory pricing policy, although successful in winning a few shippers since entering the market, may not be sustainable in the longer term, nor indeed be successful in attracting more shippers away from 'X', given 'Y's' more limited service. 'Y's' ships have a greater capacity and although they had secured some base cargo their load factor currently may not be high enough to be financially viable in the longer term.

### 3.1 Decisions facing 'X'

The decision facing 'X' is whether to take up (exercise) the option to purchase the fifth vessel or let the option lapse, and the decision has to be made now. If the vessel is not ordered by the end of the month and the competitor withdraws from the trade, then 'X' will have missed a valuable opportunity to expand its market share through its increased fleet size. The company also faces the possibility that even if 'Y' withdraws a new competitor may enter, particularly if the trade picks up suddenly. If 'X' waits for the competitor to withdraw before making its investment decision then, even if the shipyard had available capacity to begin building immediately, it would be at least a further 15 months before the new vessel could be put into service. The original deal to purchase four ships with an option on the fifth had allowed had allowed the company to negotiate an attractive price for the vessels. If 'X' does not take up the option now then this price discount may not be available at a later time. However if 'X' takes up the option of the fifth ship and the competitor makes further inroads into 'X's' market share then a fleet of five ships may not be financially viable, forcing the company to lay up, sell or charter out vessels to avoid over capacity and financial losses.

The company should appreciate that it has real options and thus flexibility to alter the project when more information is known. These options have value and should be factored into the strategic investment decision analysis, given the uncertainty of the market environment.

For this chapter we will assume that the fifth ship, if ordered (the option to build exercised), may be chartered out as an alternate strategy should the company face worsening competitive factors in 15 months time.[27] However, should market conditions improve then the company will exercise the option to switch to a five-ship service in the Trans-Tasman trade. ROA will be used to expand traditional (inflexible) DCF analysis to value this flexibility to switch operational modes (a switching option).

## 4 Basic model

The operations model is based on four 1,500 TEU ships covering five ports, three in Australia and two in New Zealand. Each ship makes 22 voyages a year.

The total market is modelled as a mean reverting random walk with a mean 300,000 TEU p.a. It is assumed to be growing by 5 per cent p.a. in line with the growth of world trade. The adjustment parameter is 4 per cent per month. The share of the market belonging to the company 'X' is 80 per cent. The standard deviation of the random disturbance is 460 TEUs per month. The total market is modelled at monthly intervals.

Load factors are limited to a maximum of 90 per cent of the ship's TEU capacity. Revenue is currently AUD1,000/TEU at start with a mean of AUD1,100. Revenue/TEU is modelled as a mean reverting random walk with a mean of AUD1,100 but declining by 1 per cent per year. The adjustment parameter is 5 per cent per month. The standard deviation of the random disturbance is AUD40 per TEU per month. Results of these parameters were validated visually by a ship owner. Revenue per TEU is modelled at three monthly intervals.

Four variable cost components were modelled: first, $550/TEU covering agency terminal costs. The second component is the cost per port visit, modelled at AUD10,000 per port visit. The third component is the ship operating costs (crew and insurances) at AUD600,000 per ship per year and the fourth is bunker costs at $228,000 per voyage. Fixed costs p.a. (net of the financing costs) are AUD700,000.

The presence of a competitor is modelled. One competitor can exist in the market at any one time. The lifetime of a competitor is modelled as an exponential distribution with a mean of one year, minimum life three months. The time between competitors is modelled as an exponential distribution with a mean of 1.5 years. When a competitor is in the market, revenue/TEU decreases by 20 per cent. The competitor's share of the market is modelled by a mean reverting random walk with expected share of 30 per cent and initial share of 20 per cent (of the main carrier's 80 per cent share). The adjustment parameter is 10 per cent per month. The standard deviation of the random disturbance is 0.02 per month.

The cost of the fifth vessel is USD20 million which at current exchange rates[28] equals AUD35,714,286. The time horizon is 16.25 years. An economic life of the vessel is assumed to be 15 years and is consistent with the tax depreciation allowance provisions, plus 15 months for the construction phase, i.e. 16.25 years.

The operations of the company were simulated over a five-year period. Terminal values were modelled as the present value of the stream of future cash flows over the life of the vessel estimated as the average cash flow of the past three of the five-year period. There were 10,000 random runs in each simulation. The output of the model is the distribution of returns over the first year.[29]

## 5  The results

The model is seeking the value of a European call option[30] to switch exchange operations from a four to a five-ship service after 15 months. Switching options are portfolios of call and put options[31] that allow their owners to switch at fixed cost (or costs) between modes of operation,[25] for example the ability of 'X' to switch between four- and five-ship services. Should market conditions not be favourable when the ship is delivered, the company would operate a four-ship service and charter out the fifth vessel. The strategy to expand operations from a four-ship to a five-ship service would only occur if the regular trade is strong.

### 5.1  Base (inflexible) case

To value the fifth ship the company performs the usual DCF analysis at 10 per cent cost of capital. The base case PV of the four-ship service is first calculated and then re-run with five ships. In recognition of the fact that there would be less competition when the company has a five-ship service, the mean time between competitors entering the market is increased from 1.5 years to 2 years. The four-ship service model assumed that there was a competitor initially in the model. Because the five-ship service model is commencing 15 months in the future, the presence of a competitor is modelled as a random event. The PV of the fifth ship is calculated as the PV of the five-ship service model discounted back 15 months (order and delivery period) less the PV of the four-ship service model. This was found to be AUD31,397,880 (AUD189,294,073 – AUD157,896,193) and is less than AUD33,802,328,[32] the present value of the vessel's building price. The NPV is thus a negative value (−AUD2,404,448). Base case (inflexible) NPV decision rules[33] would suggest that the company should forego the building of the fifth vessel, i.e. the option granted by the Korean yard would lapse.

Thus in the absence of managerial flexibility (or real options) the DCF analysis rule leads to a rejection of the project. Traditional NPV analysis assumes that management is static and ignores their ability to alter strategies in response to new information. The base case NPV (value of the project without flexibility) undervalues the project by not including in the investment decision the value of flexibility created by strategic/real options.

### 5.2 Valuing the option

Real options analysis values managerial flexibility by explicitly considering appropriate action at future dates on which information about the project's profitability is revealed. The project's true valuation may look more attractive when the value of this flexibility is incorporated.[34] This means that a project may have a static (inflexible) NPV that is negative and therefore would normally be rejected using traditional NPV rules but may be accepted if the value of flexibility is added. The static approach's assumptions do not provide a satisfactory course of action for management, should the modelled scenario not eventuate as planned. What if in 'X's' situation the vessel is ordered but the shipping line finds that in 15 months the competitive/economic market environment had not improved or had indeed worsened? 'X' would not, as the traditional approach assumes, just go on as if the modelled cash flows did exist until the end of the vessel's useful life. Obviously the company could initiate other strategies, as discussed in section 3.1.

For the purposes of this chapter, we will assume that management exercises the option with the Korean yard for the delivery of their fifth vessel in 15 months time for AUD35,714,286 and this cost will be paid on delivery.[35] However management believes that, should the competitive environment deteriorate, then, the fifth ship could be chartered out. We assume that the present value of the fifth vessel if chartered out in 15 months time will have a present value of AUD26,785,714 (USD15 million).[36] However the shipping line is keen to know the value of the exchange option created.

Scenario of section 4 was simulated and the exchange option valued by applying the recombining tree method as outlined in note 37. The value today of a flexible five-ship system with the ability to charter out, should market conditions deteriorate when the fifth ship is delivered, is AUD207,066,608. The NPV of AUD171,352,322 was calculated by deducting the discounted value of the ship (AUD33,802,378). The value of the exchange option is thus the difference between the NPV for a five-ship flexible system (AUD155,491,745) and equals AUD17,772,535.

Value of project with flexibility = Value of project without flexibility
+ Value of flexibility
AUD173,264,280 = AUD155,491,745 + AUD17,772,535

Alternately we could value the option by considering the value of the incremental fifth ship, as in the base inflexible model in Section 5.1, by taking the difference between the PV of the flexible five-ship service (AUD181,714,863) and the PV of the four-ship service (AUD157,896,193). Thus the PV of the fifth ship is AUD23,818,669. Its NPV of AUD15,368,087 is found by subtracting from this value the difference between the discounted cost of the fifth vessel (AUD33,802,328) and the PV of the charter (AUD25,351,746), i.e. AUD8,450,582. The value of flexibility (the real option) is the difference between the value of the fifth vessel without flexibility and the value of the fifth ship with flexibility. This option value of AUD17,772,835 is the same as that calculated above when valuing the project as a five-ship service.

Value of 5th ship with flexibility = Value of 5th ship without flexibility
+ Value of flexibility
AUD15,368,087 = AUD2,404,448 + AUD17,772,535

## 6 Conclusion

A real options approach radically alters the traditional way of valuing strategy. The traditional approach is to apply DCF analysis to mutually exclusive strategies and compare the outcomes. However, standard capital budgeting techniques cannot capture the value of management flexibility to respond to changes in market conditions, forcing managers to override DCF value analysis relying heavily on qualitative 'strategic' judgement (managerial experience).

ROA, and exchange options in particular, provides a better framework for valuing strategic, recognizing the fact that management does not operate in a *ceteris paribus* world. Management makes decisions under uncertainty. Once the project is underway often new information comes along indicating that it may be more appropriate to switch to a new strategy. Management must not think of strategic choice in terms of mutually exclusive scenarios but rather as a switching option exercise, providing that the cost of switching is lower than the benefit from altering course.

Shipping lines face dynamic economic/competitive environment and often need to make strategic decisions under uncertainty. A shipping line does not always have the luxury to defer investments until that uncertainty is resolved, otherwise it will not survive in the long run. However, the flexibility to alter strategies when new information becomes available can be modelled by explicitly incorporating this flexibility or real options into the investment decision.

This chapter is part of on-going research into real option analysis applied to maritime investments and uses a European call option to value changing between modes of operation, in this case between a four-ship and a five-ship service in the Trans-Tasman trade. Typical economic and competitive market conditions for a shipping line were modelled and investment scenarios valued. A base case (inflexible) four- and five-ship services investment analysis was firstly conducted and outcomes compared. The strategy to expand the service to five ships would be rejected under traditional or static DCF rules as the NPV was negative (−AUD2.404 million). This approach assumed that the four-ship and five-ship operations were mutually exclusive. This static or inflexible approach severely underestimated the value of the project by AUD17.772 million. ROA demonstrated how to value the flexibility to alter strategy by changing from a five-ship service to a four-ship service and chartering out, should market conditions worsen when the fifth vessel was delivered. However, when the flexibility to switch between four- and five-ship services was incorporated into the model, the net gain to the shipping line of AUD8.45 million demonstrates the value of the ROA approach in strategic investment analysis under uncertainty.

## References and notes

1 Bendall, H.B. and Manger, G., 1991, Corporate investment decisions: what are the relevant criteria? *Corporate Management*, 43, 3, May–June, 64–9.
2 Donaldson, G. and Lorsch, J., 1983, *Decision Making at the Top: The Shaping of Strategic Direction* (New York: Basic Books).
3 Lai, V.S. and Trigeorgis, L., 1995, The strategic capital budgeting process: a review of theories and practice, *The Capital Budgeting Process Real Options in Capital Investment: Models, Strategies and Applications*, edited by L. Trigeorgis (London: Praeger), 69–96.
4 Bendall, H.B. and Manger, G., 1988, Corporate Governance, Capital Structure and Budgeting: Preliminary Evidence from the Maritime Industry, *Inaugural Australasian Finance and Banking Conference, Australian Graduate School of Management*, Sydney, December.
5 Teisberg, E.O., 1995, Methods for evaluating capital investment decisions under uncertainty, *The Capital Budgeting Process Real Options in Capital*

*Investment: Models, Strategies and Applications*, edited by L. Trigeorgis (London: Praeger), Ch 2, 31–46.

6 An option is a financial derivative traded on exchanges or in the over the counter market (OTC). The option gives the right but not the obligation to buy or sell an underlying asset or income stream in the future. It enables the holder to benefit from upside gain while limiting downside losses to the price paid for the option, its premium.

7 Kester, W.C., 1984, Today's options for tomorrow's growth, *Harvard Business Review*, March–April, 153–60.

8 Myers, S.C., 1977, Determinants of corporate borrowing, *Journal of Financial Economics*, 5, 2, November, 147–76.

9 Moel, A. and Tufano, P., 2002, When are real options exercised: An empirical study of mine closings, *The Review of Financial Studies*, 15, 1, 35–64.

10 Bendall, H.B. and Stent, A.F., 2002, Investment decision strategies in an uncertain world: a real options approach to investments in the maritime industry, *Proceedings ICHCA 2002, 26th Biennial Conference*, Yokohama, Japan, 13–15 April, 197–218.

11 Bendall, H.B., 2002, Valuing Maritime Investments Using Real Options, *The Handbook of Maritime Economics and Business*, edited by C. Th. Grammenos (London: LLP).

12 Bjerksund, P. and Ekern, S., 1995, Contingent claims evaluation of mean reverting cash flows in shipping, *Real Options in Capital Investment. Models, Strategies and Applications*, edited by L. Trigeorgis (London: Praeger) Ch 12, 207–19.

13 Dixit, A. and Pindyck, R., 1994, *Investment Under Uncertainty* (Princeton, IL: Princeton University Press).

14 Trigeorgis, L. (ed.), 1995, *Real Options in Capital Investment. Models, Strategies and Applications* (London: Praeger).

15 Trigeorgis, L., 1996, *Real Options. Managerial Flexibility and Strategy in Resource Allocation* (Cambridge, MA: The MIT Press).

16 Black, F. and Scholes, M., 1973, The pricing of options and corporate liabilities, *Journal of Political Economy*, 81, May–June, 637–59.

17 Merton, R.C., 1973, Theory or rational option pricing, *Bell Journal of Economics and Management Science*, 4, 141–83.

18 Kasanen, E. and Trigeorgis, L., 1993, A market utility approach to investment evaluation, *European Journal of Operations Research*, Special Issue on Financial Modelling, 74, 2, April, 294–309.

19 Trigeorgis, T., 2000, Real options and financial decision-making, *Contemporary Finance Digest*, 3, 5–42.

20 Cox, J., Ross, S. and Rubenstein, M., 1979, Option pricing: a simplified approach, *Journal of Financial Economics*, 53, 1, 220–63.

21 Cox, J. and Ross, S., 1976, The valuation of options for alternative stochastic processes, *Journal of Financial Economics*, 3, 1, 145–66.

22 The risk-neutral probability approach is mathematically equivalent then a hedge portfolio is created, composed of one share in the underlying risky asset and a short position in 'm' shares of the option being priced. The hedge ratio m is riskless as loss on the underlying asset is offset by the gain on the option (and vice versa) – hence risk free.

23 Stock price of a similar (perfectly correlated) non-levered company with the same risk characteristics.

24 For example, the demand for oil may be used as a 'twin product' when valuing options associated with tankers, if volatilities are similar.

25 Copeland, T. and Antikarov, V., 2001, *Real Options* (New York: Texere).

26 Copeland and Antikarov call this assumption that the base project can be used as the marketable asset, Marketed Asset Disclaimer (MAD).

27 The company could negotiate to charter out the vessel in 15 months time with an opt out clause.

28 AUD1 = USD0.56 at time of writing.

29 Copeland, T. and Antikarov V., 2001, *Real Options* (New York: Texere).

30 See section 2.1.

31 A portfolio of American puts and calls would allow the operator to switch back and forth between four- and five-ship services. However, this is beyond the scope of this chapter.

32 See note 35 below. The cost of the vessel in 15 months should be discounted back to a value today, i.e. AUD35,714,286/1 + 0.045^1.25 where r = 4.5 per cent, the risk free rate.

33 See DCF Decision Rules in section 1.

34 However, although the value of flexibility is always positive, the price (premium) paid for the real option may exceed the additional value added to the basic NPV. In this case the NPV (with flexibility) would be negative and the investment would not go ahead. For example, a dual-fired engine system may offer the ability to switch fuel inputs but the cost of the sophisticated technology may exceed the benefit of this flexibility.

35 Generally building contracts specify that payments be made throughout the building phase. These payments relate to specified completion points. However, the assumption made in this model that payment is made on delivery in 15 months time is not of concern as a financial institution could easily arrange for these payments, secured against a letter of credit from the shipowner. This is a financing decision unrelated to the investment decision.

36 Clewlow, L. and Strickland, C., 1998, *Option Pricing: Numerical Methods* (New York: John Wiley and Sons).

37 A further development of the model will be to allow the charter rate to be stochastic.

# 4 International comparison of market risks across shipping-related industries

*Manolis G. Kavussanos, Arne Juell-Skielse and Matthew Forrest*

## 1 Introduction

A major question throughout the years has been how to finance investments in the shipping industry. Ships cost millions, and such large sums need careful investment decisions. Methods of financing have varied over time and place, as well as with the corporate structure of the company requiring funds to invest in shipping. Thus, while traditional borrowing from banks has always been prominent in the industry, charter backed finance has been very popular in the post-Second World War period. This has been followed by asset-backed finance in the 1980s (e.g. ship funds) and lately – in the 1990s – a lot of interest has been placed in drawing funds from the public. The latter is materialized either by borrowing through the issuance of bonds or by offering part-ownership to the public through shares in the company.

With respect to this last form of finance, it is of interest to potential investors and financiers to have a fair valuation of shares in the industry and to have a measure of risk-return profiles in the industry. At the same time, the listing of companies in the sector in stock exchanges around the world enables the calculation of objective market-related risks and the comparison of these with other sectors. Previous studies in the literature, such as those of Kavussanos and Marcoulis,[1-6] have concentrated on the valuation of listed companies in the US water transportation industry at the aggregate level. They find that the market risk in the industry, if not lower, is not different from that of the market and from other US transportation sectors, with the exception of rail transportation from which it has a significantly lower value. This chapter extends the work of these studies by investigating an international portfolio of shipping companies (that is, a portfolio with shares listed across stock exchanges around the world) and by comparing the risk return profiles of different sub-sectors of the shipping industry.

The CAPM (Capital Asset Pricing Model), under which the market index alone is assumed to be driving market returns, is used to estimate and compare average $\beta$s for the following, broadly-defined, sub-sectors of the shipping industry: Bulk, Container, Cruise, Drilling, Ferry, Offshore, Tanker and Yard. Such a comparison has not been attempted before and it is perceived that it would enhance one's understanding of the risk-return profiles of companies operating in these sectors, resulting in more refined investment decisions and help shipping and shipping-related companies when considering expansion and/or diversification.

The remainder of the chapter is structured as follows. The following section discusses briefly the use of the CAPM as a model of equilibrium returns and a vehicle through which to measure non-diversifiable risks. Section 3, the data section, describes how listed water transportation companies and other shipping-related stocks are classified by the industry sector depending on core business activity. The properties of the data set are also discussed here. Section 4 presents the results; average sector βs are estimated and compared with the market and between themselves to establish whether the risk/return relationship differs across sectors of the shipping industry. Finally, section 5 of the chapter concludes.

## 2 Theory – methodology

Sharpe[7] and Lintner[8] independently developed the CAPM as a general equilibrium model for asset returns. In its simplest form, it postulates that the expected return on a firm's equity can be explained as a linear function of a single factor – the expected return on the market portfolio of assets. The ideas relate back to the seminal work by Markowitz.[9,10] He developed the theoretical work, which considers an investor aiming to maximize utility derived from the returns obtained by holding individual assets. It is assumed that investors are risk averse, in the sense that for the same level of return investors would select stocks which have less risk. That is, he shows that there is an inverse relationship between stock returns and risk. Furthermore, Markowitz shows that, when investors hold more than one stock in their portfolio, the overall risk in the portfolio will decrease, thus offering safer returns to investors.

When large portfolios are held, it is not easy to calculate the relevant variance-covariance matrix, which measures the risk of these portfolios. As a consequence, Sharpe and Lintner developed a single index model, which showed that returns could be explained in terms of a single factor; the market return. This development has made the technical problem of calculating variance-covariance matrices tractable, even for large portfolios. Hence, the birth of the CAPM. At the same time, the model provides a theory, under which, if all investors behaved as if this is the model that explains stock returns, this becomes the general equilibrium model of returns in the economy.

CAPM divides the risk of holding an asset into two parts, systematic or market risk and non-systematic or specific risk. The systematic or market risk is the part related to the riskiness of the market portfolio; the non-systematic or specific risk is the residual part of the risk, which cannot be explained by the market and is company specific.

An investor can avoid the residual, non-systematic or specific, risk by holding a diversified portfolio. Accordingly, an investor should receive no added return for bearing diversifiable risk and, therefore, the expected return of an asset should only reflect the systematic or market risk. Mathematically, the CAPM shows this relationship in time series form by the following equation:

$$\tilde{R}_{it} = R_{Ft} + \beta_i(\tilde{R}_{Mt} - R_{Ft}); \ i = 1, \ldots, n; \ t = 1, \ldots, T \tag{1}$$

where $\tilde{R}_{it}$ is the expected return of stock $i$ at time $t$, $R_{Ft}$ is the risk-free rate of interest at time $t$, $\beta_i$ is the beta for stock $i$ and $\tilde{R}_{Mt}$ is the expected return from the market at time $t$.

Miller and Scholes[11] showed that equation (1) cannot be used to accurately estimate $\beta$, because $R_{Ft}$ is not constant over the estimation period. This can be shown by rearranging this equation to give:

$$\tilde{R}_{it} = (1 - \beta_i)R_{Ft} + \beta_i\tilde{R}_{Mt}; \ i = 1, \ldots, n; \ t = 1, \ldots, T \tag{2}$$

If $R_{Ft}$ fluctuates over time and if it is correlated with $\tilde{R}_{Mt}$ there is a case of missing variable bias and [gb]$_i$ will be a biased estimator of the true [gb]$_i$. Black et al.[12] solved this problem by taking as their basic time series model:

$$\text{if } R_t - R_{Ft} = \alpha_i + \beta_i(R_{Mt} - R_{Ft}) + e_{it}; \ i = 1, \ldots, n; \ t = 1, \ldots, T \tag{3}$$

where $R_{it}$ is the holding period return on the equity of company $i$ in period $t$, $R_{Ft}$ is the risk free rate, $R_{Mt}$ is the holding period return on the market portfolio of stocks in period $t$, when $e_{it}$ is the residual left unexplained – the non-systematic or specific risk.

$\alpha_i$ and $\beta_i$ are the CAPM parameters for stock $i$. The $\alpha$ indicates whether the stock is trading at a fair price. The $\beta$ is a measure of the stock's sensitivity to changes in the expected market return. The CAPM suggests that an average stock would have a $\beta$ value of one and, if correctly priced, an $\alpha$ of zero. A negative $\alpha$ indicates that the stock is overpriced, since its return is higher than that implied by the CAPM; a positive $\alpha$ indicates that the stock is underpriced, since its return is lower than that implied by the CAPM. A stock with a $\beta$ greater than one carries above average systematic risk and an investor would, therefore, require a higher expected return to hold it. Conversely, a stock with a $\beta$ less than one carries below average systematic risk.

Equation (3) is estimated for each company by Ordinary Least Squares (OLS) to obtain estimates for $\alpha$ and $\beta$, say $\hat{\alpha}$ and $\hat{\beta}$, respectively. Averages of these, denoted $\bar{\hat{\alpha}}$ and $\bar{\hat{\beta}}$, respectively, are then calculated for each sector, and are used as estimates of the true sector $\bar{\alpha}$ and $\bar{\beta}$s. Two separate null hypotheses are tested then; that is that $\bar{\alpha} = 0$ and that $\bar{\beta} = 1$ in order to test for underpricing and whether the sector average $\beta$ is 1, respectively. The following test statistics are used:

$$\frac{\bar{\alpha}_i - \bar{\alpha}}{SE(\bar{\alpha}_i)} \sim t(n_i - 1) \quad \text{and} \quad \frac{\bar{\beta}_i - \bar{\beta}}{SE(\bar{\alpha}_i)} \sim t(n_i - 1) \tag{4}$$

where $\bar{\alpha}_i$, $\bar{\beta}_i$, $\bar{\alpha}$ and $\bar{\beta}$ are the sector $i$ averages of the estimated parameters and the values of the null hypotheses being tested for, and $SE(\bar{\alpha}_i)$ and $SE(\bar{\beta}_i)$ are the estimated standard errors of these averages.

The average $\beta$s of each sub-sector are also compared to examine whether systematic risks differ between sub-sectors of the international[13] water transportation industry. The test statistic used is:

$$\frac{\bar{\beta}_i - \bar{\beta}_j}{\sqrt{\dfrac{S^2}{n_i} + \dfrac{S^2}{n_j}}} \sim t(n_i + n_j - 2) : i \neq j \quad \text{where} \quad S = \frac{(n_i - 1)S_i^2 + (n_j - 1)S_j^2}{n_i + n_j - 2} \tag{5}$$

where $\bar{\beta}_i$, $\bar{\beta}_j$, $S_i$, $S_j$ and $n_i$, $n_j$ are the sector $i$ and $j$ averages of the estimated parameters, their standard deviations and the number of companies in sector $i$ and sector $j$, respectively. $S$ is a weighted average of $S_i$ and $S_j$, which is used because the sample sizes are small.

The theoretical underpinning behind the above procedure is that the risk-return profiles of stocks in the economy vary according to the industry they belong to. Earlier work at the industry level[14-16] has shown that firms within the same industry experience similar rates of return. Furthermore, industry average rates of return exhibit significant differences; and this is reflected in the increasing focus of investors towards an industry-oriented approach by the existence of sectoral funds (e.g. transport, construction, banking, etc.).

Furthermore, studies such as Capaul[17] and Weiss[18] argue that, as capital market integration develops and certain global industries are to a certain degree homogeneous, the industrial classification of a given asset becomes increasingly important to the investor. This raises the need to study industries at the international level – across country borders. That is, it is argued that the degree of integration in various industries is such, at the international level, that global asset management firms increasingly place an industry focus in their research. In addition, the world economy is becoming increasingly more globalized, with companies operating across borders, forming alliances/mergers in several industrial sectors, in some of them more than others. Given this industry internationalization in the world economy, the current study fits well within this framework, particularly in an international industry such as shipping.

## 3 Data classification and properties

### *3.1 Classification of companies by maritime sector*

For analysis, this chapter identifies every possible maritime company listed continuously in any stock exchange in the world over the most recent three-year period and classifies it under pre-defined sub-sectors of the industry (see Table 4.1 for details). This, sampling of companies across stock exchanges (rather than focusing on companies listed in one exchange) gives the largest possible cross-sectional sample of maritime companies in each sub-sector, and at the same time a sufficient length of time series data for returns (36 monthly observations) to enable estimation and inferences.

The starting point is the Maritime Transport and Energy list of traded shares appearing in the 'Financial World' page of the *Lloyds List*. This is supplemented by any other public companies known to be involved in shipping or shipping-related industries but not listed there.[19,20] In order to classify companies into sectors, a short questionnaire was sent to 250 of these companies in July 1999, asking them to classify the percentage of their core business activity in a number of pre-defined sectors. This information was supplemented by consulting their annual reports for 1998 and 1995. There was an approximately 20 per cent response to the initial questionnaire, with a further 30 per cent replying after a reminder letter, which was sent four weeks later. Financial information for companies that

*Table 4.1* Maritime industry sectors

| Sector | Description |
| --- | --- |
| Bulk | Dry bulk, older type General cargo ships, excluding OBO. |
| Container | LOLO and some ROLOs with large container section. |
| Cruise | Cruise ships. |
| Drilling | Rig owners and operators. |
| Ferry | Passenger ferries including ROPAX. |
| Offshore | Supply boats and anchor handlers. |
| Shipping | Companies with 90% or more of revenue derived from shipping or shipping-related activities but which could not be classified into any other sector. |
| Tanker | Oil Tanker, excluding chemical and gas Tankers as well as FSPO. OBOs were included when operated as oil Tankers. |
| Yard | Shipyards excluding Rig yards. |
| Diversified | Companies with between 60–90% of revenue derived from shipping or shipping related activities – the balance being derived from elsewhere. |
| All | All of the above sectors. |

Notes
OBO: Oil Bulk Ore; LOLO: Lift On Lift Off; ROLO: Roll On Lift Off; ROPAX: Roll On Passenger.

did not reply was obtained from the Wright Investors' Service web page (http://www.wisi.com), from the Fairplay Online Directory (http://www. wsdonline.com) and from individual company web pages.

In order to make inferences for each sector which reflect the risk-return profile of operating in the specific sector, companies whose economic activity in shipping or shipping-related activities was less than 60 per cent were considered overly diversified and were discarded from the sample.[21] Companies for which there was no information on revenue, companies involved in mergers, acquisitions and/or changes in their core business during the sample timeframe and companies for which stock data could not be found on DataStream were excluded from the analysis.

To account for the possibility that different degrees of diversification have varying effects on the risk-return profiles of sectors, the companies that remained were classified and analysed according to whether 60, 75 and 90 per cent of their core business activity was in the same sector. Specialized companies operating only in one sector were straightforward to classify. Companies whose core activity was over 90 per cent in more than one sector of shipping but for which no detailed breakdown of the percentages attributable to each sector were available were classified in a general category called 'shipping'. Companies with diverse core business that included over 10 per cent of activities not shipping-related were classified as 'diversified'. The sectors 'Reefer', 'Gas', 'Chemical Tankers', 'Brokers' and 'Ports' had to be abandoned due to too few listed companies belonging to them. In total, 108 companies made up the final sample used for analysis. The number of companies in each sector under the 60, 75 and 90 per cent classification criteria are shown in Table 4.2.

### 3.2 Data sources and summary statistics

Monthly stock price and dividend yield (in percentage form) data for each share are collected from DataStream International Service. Logarithmic monthly returns for company $i$ at time $t$, $R_{it}$, are calculated in percentage form using the equation:

$$R_{it} = 100 * \ln\left[ \frac{(P_{it} + (P_{it} * DY_{it}/1200))}{P_{it-1}} \right] \quad (6)$$

where $P_{it}$ and $P_{it-1}$ are the stock prices of company $i$ at time $t$ and $t-1$, respectively, and $DY_{it}$ is the annualized dividend yield paid by company $i$ at time $t$.

In calculating the CAPM regression of equation (3), a question of what is the relevant market is always raised. Because the sample includes companies listed on stock exchanges in different countries, the Morgan Stanley Capital International (MSCI) All Country World Index was used for analysis. Given the recent developments of the launch of maritime funds

Table 4.2 Summary statistics of mean monthly returns of each sector by classification criteria; July 1996–July 1999

| Sector | Classification criteria | | | | | | | | | Skew | Kurt |
|---|---|---|---|---|---|---|---|---|---|---|---|
| | 90% | | | 75% | | | 60% | | | | |
| | Mean | SD | No | Mean | SD | No | Mean | SD | No | | |
| Bulk | -2.18 | 1.22 | 6 | -1.79 | 1.54 | 7 | -1.88 | 1.45 | 8 | – | – |
| Container | -0.85 | 2.90 | 7 | -0.92 | 2.45 | 9 | -0.92 | 2.45 | 9 | – | – |
| Cruise | 3.04 | 1.32 | 3 | 3.04 | 1.32 | 3 | 2.93 | 1.10 | 4 | – | – |
| Drilling | 0.32 | 1.12 | 7 | 0.32 | 1.12 | 8 | 0.33 | 1.05 | 9 | – | – |
| Ferry | -0.05 | 2.73 | 11 | -0.47 | 2.60 | 15 | -0.14 | 2.61 | 17 | – | – |
| Offshore | 0.17 | 1.67 | 7 | 0.17 | 1.54 | 8 | 0.25 | 1.38 | 10 | – | – |
| Shipping | -1.68 | 3.79 | 34 | -1.77 | 3.90 | 30 | -2.01 | 3.76 | 30 | – | – |
| Tanker | -2.53 | 2.50 | 12 | -2.53 | 2.50 | 12 | -2.46 | 2.41 | 13 | – | – |
| Yard | 0.60 | 0.46 | 4 | 0.23 | 1.10 | 6 | 0.23 | 1.00 | 7 | – | – |
| Diversified | -0.50 | 1.57 | 17 | -0.12 | 1.72 | 10 | N/A | N/A | N/A | – | – |
| All | -0.92 | 2.85 | 108 | -0.91 | 2.86 | 108 | -0.91 | 2.86 | 107 | – | – |
| MSCI-All | 1.42 | 4.50 | – | – | – | – | – | – | – | -1.61 | 4.19 |
| MSCI-Sh | 0.28 | 6.34 | – | – | – | – | – | – | – | 4.19 | 3.65 |

Notes
SD = Standard Deviation, No = Number of companies classified under each sub-sector, Skew = Coefficient of Skewness, Kurt = Coefficient of Kurtosis,
MSCI-All and MSCI-Sh are the Morgan Stanley All Country World Index and the Shipping Index, respectively.
Under the 60% criterion, the Diversified sector only contained one company (Wilh Wilhelmsen) and this sector was, therefore, dropped.

and the practice of evaluating industry-specific funds by benchmarking on sectoral indices, the MSCI International Shipping Index was also used for analysis.

The MSCI All Country World Index is calculated as a market capitalization weighted average of equity returns in 51 countries (23 developed and 28 emerging) and is quoted in gross form inclusive of dividends. The MSCI Shipping Index is one of the 38 industry indices produced by Morgan Stanley. Companies are classified based on their principal economic activity, as determined by the breakdown of earnings, which is in line with the classification method of maritime companies into sectors in this chapter. If no detailed earnings data are available, then breakdown of sales data are used. In defining industries, MSCI attempts to construct homogeneous groups which are expected to react similarly to economic and political trends and events. Logarithmic monthly market percentage returns, $R_{Mt}$, are calculated for both the MSCI All Country World Index and the MSCI Shipping Index using equation (6), with the dividend part being excluded. The US three-month Treasury bill rate is used as a measure of the global risk free rate of interest, $R_{Ft}$. Table 4.2 presents summary statistics for average equity returns by the maritime sector and for the returns on the MSCI world and shipping indices for the period July 1996 to July 1999.

Both the MSCI All Country World Index and the Shipping Index have positive average monthly returns over the sample timeframe; however the Shipping Index under-performs the All Country World Index and has higher volatility than the latter. Both indices have negatively skewed and leptokurtic distributions. Turning next to returns in maritime sectors, it can be seen that most sectors performed poorly over the same period, with six out of the ten sectors showing negative average monthly returns, including the 'all' sector category.

The tanker sector seems to be the worst performer, followed by Bulk, Shipping and Container. Figures 4.1, 4.2 and 4.3 give an idea of how revenues in the Tanker, Bulk and Container markets faired over the sample timeframe. Although all three performed well up until early 1997, they declined steadily after that. Therefore, although alarming, the negative average monthly returns are, nonetheless, not unexpected. The Cruise sector appears to be the best performer. However, it should be noted that this sector has the fewest members – only four companies with the 60 per cent classification criteria and three at the 75 and 90 per cent criteria. Nevertheless, it is also true that this sector has enjoyed considerable prosperity and growth over the sample timeframe and, therefore, this result is again not unexpected. Also showing positive average monthly returns are the Yard, Drilling and Offshore sectors.

The fact that the Yard sector has enjoyed positive returns whilst most other shipping sectors have suffered negative returns is not totally unexpected. If the sample timeframe is considered as the 'collapse' phase of the

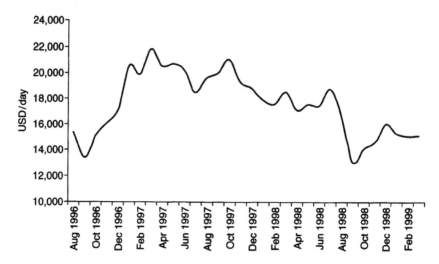

*Figure 4.1* Weighted average tanker earnings (source: Clarkson Research).

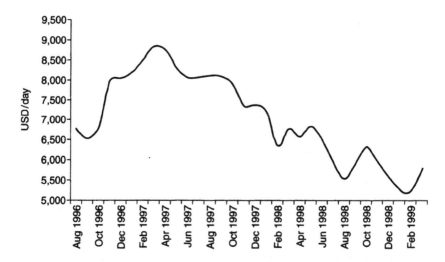

*Figure 4.2* Weighted average bulk carrier earnings (source: Clarkson Research).

shipping cycle, it is not unreasonable to expect that yards would be busy completing orders placed during the previous 'peak' of the cycle. The Drilling and Offshore sectors are probably more correlated with the oil price than with other shipping sectors. Whilst both negative, the average monthly return of the Diversified sector (which contains companies with up to 40 per cent of revenue attributable to non-shipping or non-shipping-related activities) is higher than that of the Shipping sector (which con-

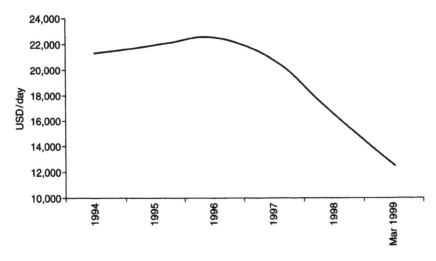

*Figure 4.3* Average one-year T/C freight rates FCC 2,750 TEU (source: Clarkson Research).

tains companies with over 90 per cent of revenue attributable to diverse shipping or shipping-related activities). This would suggest that companies that diversified outside of shipping or shipping-related activities performed better over the sample timeframe.

All sectors have lower total risk (SD) than the MSCI world index. The Shipping sector exhibits the highest total risk, followed by Ferry and then Tanker. Interestingly, sectors with positive average monthly returns generally exhibit relatively low total risk when compared to sectors with average monthly returns that are negative. However, the Bulk sector is the exception to this, having one of the highest negative average monthly returns and relatively low total risk. The Yard sector has the lowest total risk followed by the Drilling sector.

## 4 Systematic risk and CAPM results

As company-specific risks may be diversified through portfolio formation, one should consider market or systematic risks ($\beta$), rather than total risk (standard deviation) as the metric of risk for each sector. The CAPM of equation (3) provides the tool to measure these. Tables 4.3 and 4.4 show the average (over companies in each sector) CAPM parameters estimated for each sector across classification criteria (90, 75 and 60 per cent), together with their standard errors in brackets and average $R^2$ values for the regression of excess stock returns against the excess return over the MSCI All Country World and Shipping Indices, respectively.

The $R^2$ values range from 0.02–0.35, indicating that little of the stocks'

Table 4.3 Average sector CAPM parameters across classification criteria for regression against MSCI All Country World Index; July 1996–July 1999

| Sector | Classification criteria | | | | | | | | |
|---|---|---|---|---|---|---|---|---|---|
| | 90% | | | 75% | | | 60% | | |
| | α | β | $R^2$ | α | β | $R^2$ | α | β | $R^2$ |
| Bulk | -3.07** (0.65) | 0.46* (0.19) | 0.02 | -2.87** (0.58) | 0.66** (0.26) | 0.08 | -2.98** (0.52) | 0.68** (0.22) | 0.06 |
| Container | -2.04 (1.11) | 0.76** (0.25) | 0.10 | -2.07** (0.81) | 0.73** (0.17) | 0.08 | -2.07** (0.81) | 0.73** (0.17) | 0.08 |
| Cruise | 1.63 (0.94) | 0.99** (0.17) | 0.17 | 1.63 (0.94) | 0.99** (0.17) | 0.17 | 1.72* (0.67) | 0.78** (0.24) | 0.13 |
| Drilling | -1.45** (0.45) | 1.33** (0.15) | 0.12 | -1.45** (0.42) | 1.33** (0.14) | 0.12 | -1.42** (0.37) | 1.32** (0.12) | 0.12 |
| Ferry | -1.13 (0.86) | 0.65** (0.15) | 0.08 | -1.56** (0.72) | 0.67** (0.16) | 0.07 | -1.19 (0.69) | 0.63** (0.15) | 0.08 |
| Offshore | -1.51** (0.56) | 1.25** (0.22) | 0.17 | -1.53** (0.49) | 1.27** (0.19) | 0.17 | -1.50** (0.39) | 1.32** (0.15) | 0.17 |
| Shipping | -2.78** (0.66) | 0.67** (0.13) | 0.07 | -2.83** (0.73) | 0.64** (0.14) | 0.07 | -3.09** (0.70) | 0.66** (0.13) | 0.06 |
| Tanker | -3.47** (0.79) | 0.52** (0.18) | 0.07 | -3.47** (0.79) | 0.52** (0.18) | 0.07 | -3.31** (0.75) | 0.43** (0.19) | 0.07 |
| Yard | -0.08 (0.36) | 0.26 (0.28) | 0.03 | -0.53 (0.56) | 0.34 (0.19) | 0.03 | -0.75 (0.52) | 0.56 (0.28) | 0.05 |
| Diversified | -1.74** (0.43) | -0.81** (0.17) | 0.08 | -1.37** (0.59) | 0.83** (0.22) | 0.09 | – | – | – |
| All | -2.10** (0.28) | 0.75** (0.06) | 0.10 | -2.08** (0.28) | 0.75** (0.06) | 0.10 | -2.08** (0.28) | 0.75** (0.06) | 0.09 |

Notes
Figures in brackets are standard errors.
* and ** indicate significance at the 10% and 5% levels, respectively.

*Table 4.4* Average sector CAPM parameters across classification criteria for regression against MSCI Shipping Index; July 1996–July 1999

| Sector | Classification criteria | | | | | | | | |
| | 90% | | | 75% | | | 60% | | |
| | $\alpha$ | $\beta$ | $R^2$ | $\alpha$ | $\beta$ | $R^2$ | $\alpha$ | $\beta$ | $R^2$ |
|---|---|---|---|---|---|---|---|---|---|
| Bulk | -2.56** (0.50) | 0.48** (0.12) | 0.06 | -2.15** (0.59) | 0.61** (0.17) | 0.10 | -2.24** (0.52) | 0.64** (0.15) | 0.11 |
| Container | -1.33 (1.07) | 0.68** (0.18) | 0.13 | -1.29 (0.82) | 0.54** (0.17) | 0.10 | -1.29 (0.82) | 0.54** (0.17) | 0.10 |
| Cruise | 2.67* (0.75) | 0.53 (0.19) | 0.08 | 2.67* (0.75) | 0.53 (0.19) | 0.08 | 2.55** (0.54) | 0.43* (0.17) | 0.07 |
| Drilling | 0.03 (0.40) | 1.59** (0.09) | 0.35 | 0.03 (0.40) | 1.59** (0.09) | 0.35 | 0.04 (0.35) | 1.58** (0.08) | 0.35 |
| Ferry | -0.16 (0.86) | 0.51** (0.12) | 0.10 | -0.84 (0.67) | 0.56** (0.13) | 0.10 | -0.52 (0.63) | 0.52** (0.12) | 0.09 |
| Offshore | -0.14 (0.63) | 1.30** (0.25) | 0.30 | -0.14 (0.55) | 1.35** (0.22) | 0.32 | -0.05 (0.44) | 1.37** (0.17) | 0.33 |
| Shipping | -2.04** (0.64) | 0.59** (0.09) | 0.12 | -2.14** (0.71) | 0.60** (0.09) | 0.12 | -2.38** (0.69) | 0.61** (0.09) | 0.12 |
| Tanker | -2.90** (0.72) | 0.63** (0.11) | 0.11 | -2.90** (0.72) | 0.63** (0.11) | 0.11 | -2.82** (0.66) | 0.59** (0.11) | 0.10 |
| Yard | 0.21 (0.23) | 0.33 (0.16) | 0.04 | -0.16 (0.44) | 0.34* (0.15) | 0.04 | -0.15 (0.37) | 0.45** (0.17) | 0.05 |
| Diversified | -0.85** (0.38) | 0.74** (0.15) | 0.16 | -0.47 (0.54) | 0.78** (0.20) | 0.19 | – | – | – |
| All | -1.27** (0.28) | 0.72** (0.06) | 0.14 | -1.27** (0.28) | 0.72** (0.06) | 0.14 | -1.27** (0.10) | 0.72** (0.04) | 0.14 |

Notes
Figures in brackets are standard errors.
* and ** indicate significance at the 10% and 5% levels, respectively.

behaviour is explained by the MSCI Indices. $R^2$ values in Table 4.4 are higher than those in Table 4.3 – signifying that the MSCI Shipping Index explains more of the stocks' behaviour than the MSCI All Country World Index. The only exception is the Cruise sector, which has a lower $R^2$, indicating that the behaviour of the stocks in this sector is explained more by the MSCI All Country World Index than by the MSCI Shipping Index. This is a reasonable result given the dependence of this sector on disposable incomes and the tourist industry.

Turning next to the values of the average αs, which indicate possible mispricing of stocks in each sector when they are different from zero, one observes the following: When the MSCI All Country World Index is used in the CAPM regression, all sectors have a negative α (with the exception of Cruise). Examining statistical significance, the α for the Container (under the 90 per cent criterion), Cruise, Ferry (under the 90 and 60 per cent criteria) and Yard sectors are statistically zero, indicating correct pricing of stocks. The other sectors (Bulk, Drilling, Offshore, Shipping, Tanker and Diversified) all have significantly negative α values, implying overpricing. This is also the case for the 'All' sector, which considers all the maritime sectors together.

When regressed against the MSCI Shipping Index, significance tests show that the αs for the Container, Cruise, Drilling, Ferry, Offshore and Yard sectors are statistically zero, implying that these sectors are in fact correctly priced. The Bulk, Shipping and Tanker sectors have significantly negative α values, indicating overpricing. The overall average of all the α values is also significantly negative, implying overpricing of maritime stocks. Thus, more sectors appear to be correctly priced when the shipping index is used as the basis for estimation in comparison to the All share index. It seems then that the choice of the index can lead to different results regarding the question of fair pricing of stocks and may affect the evaluation of maritime fund managers' performance as a consequence.

Turning next to the values of the β coefficients obtained when the All Country Index is used in the CAPM regression, one observes that all sectors, with the exception of Yard (and Cruise when the Shipping Index is used), have β values that are significantly different from zero (one other exception is the Bulk sector under the 90 per cent criterion). Overall, the Drilling sector has the highest β value (1.33), closely followed by the Offshore (1.25) sector for all three classification criteria. This makes sense because these are the sectors that show the highest average monthly returns and, therefore, would be expected to have the highest market risk. The other sectors have β values which are numerically smaller than one. For the 90 per cent classification criterion, the Yard sector has the lowest β value (0.26), followed by the Bulk sector (0.46) and then the Tanker (0.52), Ferry (0.65), Shipping (0.67) and Container (0.76) sectors when the all market index is used, whilst, when the shipping index is utilised, the Yard sector still has the lowest β value (albeit higher, 0.33), followed by the Bulk (0.48) and Ferry (0.51) sectors.

Across classification criteria, the sector β values show no clear pattern. As the percentage of the core activity required for classification in a sector increases, some sector β values increase, some decrease and some remain the same. For instance, the Container, Cruise, Drilling and Tanker sectors have β values which increase with increasing specialization. The Bulk, Offshore and Yard sectors have β values that decrease, while the β for the Ferry sector increases then decreases and the β for the Shipping sector does the reverse. The results are similar when the Shipping Index is used to estimate the CAPM. It seems then that the extent of diversification/specialization of maritime companies can either increase or decrease market risk according to the sector being investigated. The analysis in this chapter then helps to identify differences in market risks between sectors and how they change as companies specialize or diversify within each sector.

An interesting question for stock selection in portfolio formation is the comparison of the β values for each sector with that of the market (one), as it would indicate whether sectors carry above or below average market risk. Results show that only the Drilling sector has β values that are significantly higher than one for both the All share and the shipping indices, except when the 90 per cent classification criterion is used in the All share index. The Container (only for the 90 per cent criterion when the Shipping Index is used), Cruise, Diversified and Offshore sectors have β values that are not statistically different from one, as does the Bulk sector (except when classified under the 90 per cent criterion for the All share index and the 75 per cent criterion for the Shipping index, when it is less than one), implying that these sectors exhibit average market risk. This indicates that these sectors have the same risk as the market. For the Shipping, Tanker and Ferry (except for the 75 per cent criterion in the All share index) sectors, the β value is significantly lower than one, signifying that these sectors exhibit less than market risk. Also, the Bulk sector, as mentioned above, shows some evidence of below average systematic risk and so does the Container sector when using the Shipping Index and under both the 75 and 60 per cent criteria. The yard sector results are unreliable due to the small number of companies, which results in large standard errors; as a consequence, both the nulls of 0 and 1 are not rejected. Finally, for the 'All' sector, bundling all maritime sectors together, the β values are significantly lower than unity, implying that on average maritime stocks exhibit below average market risk.

This is in line with the results reported for the US water transportation stocks in Kavussanos and Marcoulis,[6] where they find evidence of both equal and below average market risk for the sector. It seems that the formation of international portfolios – covering stocks listed across country borders – in the industry makes the result of below average market risk more of a certainty and is in line with what one would expect *a priori* from international portfolio diversification.

Equation (5) is used to test whether sector β values differ significantly from each other. Broadly speaking, the results for both the MSCI All and Shipping Indices are as follows: For the 60 and 75 per cent classification criteria, the Drilling and Offshore sectors have average β values which are significantly higher than all the other sectors but not significantly different from each other. For the 90 per cent classification criterion, when the All Country Index is used, the Drilling and Offshore sectors cease to be significantly different from the Cruise sector, the Offshore sector ceases to be significantly different from the Container and Diversified sectors, while the Drilling sector continues to be significantly higher than these two sectors.

## 5 Further discussion

From the comparison of sector β values, for both the regression against the MSCI All Country World Index and the MSCI Shipping Index, it is clear that the Drilling and Offshore sectors have average β values that are consistently higher than all other sector β values, but not significantly different from each other. This suggests that these sectors exhibit a higher degree of market risk than all the other sectors. If this is the case, it should be expected that, on average, these sectors produce the highest returns. However, this is not found to be true, as the Cruise sector produces the highest average monthly return, followed by the Yard sector.

Also, while the average β values for the Drilling and Offshore sectors seem to be significantly different from all other sectors, only the Drilling sector average β is significantly greater than the market β value of one. This suggests that the Drilling sector has the highest risk and it should, therefore, have the highest return. Again, this does not seem to be true, as the Cruise sector shows the highest average monthly return.

When regressed against the MSCI All Country World Index, the β value for the Cruise sector becomes insignificantly different from the Drilling and Offshore sectors, as the companies that make up the sectors become more specialized (i.e. as the sector classification criterion increases from 60 to 75 to 90 per cent). This could be interpreted as suggesting that the Cruise sector does in fact exhibit more risk than the other sectors (except Drilling and Offshore), even though its β value remains insignificantly different from the β values of these sectors.

It is, therefore, apparent that the Drilling, Offshore and Cruise sectors exhibit different risk-return characteristics than the other sectors. It may be possible to explain this difference by considering the market fundamentals of these sectors together with their supply/demand characteristics. Demand for the Offshore and Drilling sectors is influenced, amongst other things, by the price of crude oil and natural gas. Even if it can be argued that the crude oil price is set by politics in the short run, it is set by demand – the world economy – in the long run. Consequently, it is reasonable to assume that these two sectors are probably more correlated with the world economy than with the shipping industry. Indeed, this argument is given further weight when one considers that these two sectors have the highest $R^2$ values of all sectors for the regressions against the MSCI All Country World Index.

It could also be argued that the demand for the Cruise sector is also more correlated with the world economy than with the shipping industry. If the world economy is prosperous then disposable income should be high, leading to an increase in the demand for tourism and travel. Certainly, the fact that the $R^2$ value for the Cruise sector is higher for the regression against the MSCI All Country World Index than for the regression against the MSCI Shipping Index implies that this sector is indeed more correlated with the world economy than with the shipping industry.

It is interesting that no significant difference could be found in the β values for the Bulk, Tanker, Container and Ferry sectors for either the regression against the MSCI All Country World Index or the MSCI Shipping Index. All these sectors have higher $R^2$ values when regressed against the MSCI Shipping Index than when regressed against the MSCI All Country World Index. This means that more of the behaviour of the stocks in these sectors is explained by what is happening in the shipping industry than by the world economy. This is as expected, although it should be remembered that the performance of the shipping industry itself is probably dependent on the state of the world economy in any case.

The β values for the Tanker and Bulk sectors were found to be consistently significantly less than one when estimated both from the regressions which used the MSCI All Country World Index and the MSCI Shipping Index. This suggests that these two sectors exhibit relatively low levels of market risk – a result that seems odd given that all these sectors showed negative average monthly returns. However, the low $R^2$ values for these sectors show that there is little correlation between the stocks in these sectors and the world economy or the shipping industry. This first statement is perhaps not surprising when one considers that the lagged delivery of new tonnage can result in these markets moving out of phase with the world economy driving them. However, one would expect these sectors to be more highly correlated with the shipping industry in general.

The result for the Ferry sector is similar to that for the Bulk and Tanker sectors. Its β value is found to be consistently significantly less than one both when extracted from the regressions against the MSCI All Country World Index and the MSCI Shipping Index. On one level it could be argued that the demand for the Ferry sector should be correlated with the demand for the Cruise sector given that both are dependent on the demand for tourism and travel. However, the Ferry sector also derives revenue from freight and in this sense it is also reasonable to argue that the risk-return profile of this sector should be more like that of the Bulk, Tanker and Container sectors. Furthermore, the Ferry sector is probably at a more mature stage in its market cycle than the Cruise sector and, as such, has a more stable demand. This argument is corroborated by the fact that the Cruise sector has been going through a period of rapid expansion over the last few years, something that has not been evident in the Ferry sector.

When using the MSCI All Country World Index in the regression, the amount of risk exhibited by the Container sector was found to be insignificantly different from that of the market. This was also found to be the case when the sector was classified using the 90 per cent classification criterion and stock returns were regressed against the MSCI Shipping Index. This suggests that the Container sector may exhibit more risk than the Ferry and Tanker sectors (because these sectors were found to exhibit significantly less than market risk for both regressions), although no significant difference was found when comparing the β values of the sectors directly. This result is hard to explain – it could be argued that the structure of the Container market should make it relatively low risk. Certainly conferences and alliances, cartels and tariffs act as barriers to entry and make the Container sector anything but a perfect market. Again, the lagged delivery of new tonnage may explain why this sector is moving out of phase with the market. Generally, the fact that the sample size was so small for this sector means that the results should be viewed with a degree of caution.

The Yard sector exhibits a wide variety of results. Again, this may be explained by the small sample and that some of the companies in the sample were favoured by government subsidies over the period, thus interfering with market factors. Orders placed during prosperous periods help this sector during periods of recession.

No significant difference could be found between the β of the Shipping sector and the β of the Diversified sector. This suggests that there is no difference in market risk for shipping or shipping-related companies that diversify their activities within shipping or outside shipping. However, although there is no significant difference between them, the β for the Shipping sector was found to be significantly different from the market for both regressions, but the β for the Diversified sector was not. Looking at the results in this way, and given that the β value for the Diversified sector is numerically higher than that for the Shipping sector, it could be argued that the Diversified sector does exhibit more market risk. This seems logical considering that the average monthly return of the Diversified sector was higher than that of the Shipping sector. This also makes sense given that this study has found that, on average, shipping stocks exhibit significantly less than average market risk.

## 6 Conclusions

The aim of this chapter has been to investigate the risk-return profiles of sub-sectors of the international shipping industry. Replies from an extensive questionnaire survey, regarding core business activities of public companies in the industry, have been supplemented with annual report and company website information to classify companies into sub-sectors. Three classification criteria (90, 75 and 60 per cent) were used to that effect, in order to identify possible differences in the risk-return profiles of each sector as the degree of diversification changed. Both an All Country World Index and a World Shipping Index have been used in the analysis.

During the 1996–99 period analysed, when the shipping industry was not doing particularly well, companies in sectors were broadly overpriced and average returns seemed to be negative. Market βs for all the stocks in the industry appeared to be significantly lower than the market. The Drilling and the Offshore sectors were significantly higher than one, however all other average sector βs appeared to be either equal or lower than the market average. The sectors that appeared to have βs which were significantly lower than the market are the Shipping, Tanker, Ferry and also Bulk and Containers mostly. It seems then that the maritime industry stocks do not carry above average market risk, at the international setting.

In comparing the βs amongst sectors, it seems that the Drilling and Offshore sectors have the same proportion of systemic risk in them. The β values of these sectors do not differ significantly from each other but are significantly different from all the other sector β values except for Cruise. However, the Cruise sector β, whilst not significantly different from Drilling and Offshore, is not significantly different from any other sector. There is no significant difference in the β values of the remaining sectors (Bulk, Container, Ferry, Shipping, Tanker, Yard, Diversified and All). When regressed against the MSCI Shipping Index, the Drilling and Offshore sectors again appear to have the same degree of market risk in them. There is no significant difference in the β values of all other remaining sectors.

Finally, as more companies in the industry become public the scope for increased sample sizes for each sub-sector of the maritime industry on which to base inferences will also increase. Perhaps a further study when more data is available and also when market conditions are different (on the upturn) may add to the body of knowledge established with this chapter.

**Acknowledgement**

The authors would like to thank all the companies that have replied to the questionnaire, thus making the research in this chapter possible. Thanks are also due to participants for their comments, in an earlier version of this chapter, presented at the International Association of Maritime Economists Conference, IAME 2001, Hong Kong.

## References and notes

1 Kavussanos, M.G. and Marcoulis, S., 1997, Risk and return of US water transportation stocks over time and over bull and bear market conditions, *Maritime Policy & Management*, 24, 145–58.

2 Kavussanos, M.G. and Marcoulis, S., 1997, The stock market perception of industry risk and microeconomic factors: the case of the US water transportation industry versus other industries, *Transportation Research, Part E*, 33, 147–58.

3 Kavussanos, M.G. and Marcoulis, S., 1998, Beta comparisons across industries – a water transportation industry perspective, *Maritime Policy & Management*, 25, 175–84.

4 Kavussanos, M.G. and Marcoulis, S., 2000, The stock market perception of industry and macroeconomic factors: the case of the US water transportation industry versus other transport industries, *International Journal of Maritime Economics*, 2, 235–56.

5 Kavussanos, M.G. and Marcoulis, S., 2000, The stock market perception of industry risk through the use of a multifactor model. *International Journal of Transport Economics*, XXVII, 77–98.

6 Kavussanos, M.G. and Marcoulis, S., 2001, *Risk and Return in Transportation and other US and Global Industries* (The Netherlands: Kluwer Academic Publishers).

7 Sharpe, W., 1964, Capital asset prices: a theory of market equilibrium under conditions of risk, *Journal of Finance*, 19, 3, 425–42.

8 Lintner, J., 1965, The valuation of risk assets and the selection of risky investments in stock portfolios and capital budgets, *Review of Economics and Statistics*, 47, 13–37.

9 Markowitz, H.M., 1952, Portfolio selection, *Journal of Finance*, 1, 77–91.

10 Markowitz, H.M., 1959, *Portfolio Selection: Efficient Diversification of Investments* (New York: Wiley). Also, 1991 (Cambridge, MA: Basil Blackwell).

11 Miller, M.H. and Scholes, M., 1972, Rates of return in relation to risk: a re-examination of some recent findings, *Studies in the Theory of Capital Markets*, edited by M.C. Jensen (New York: Praeger).

12 Black, F., Jensen, M.C. and Scholes, M, 1972, The capital asset pricing model: some empirical tests, *Studies in the Theory of Capital Markets*, edited by M.C. Jensen (New York: Praeger).

13 That is, of companies in the sector listed in stock exchanges around the world, internationally.

14 King, B., 1966, Market and industry factors in stock price behaviour, *Journal of Business*, January, 39, 1, 139–90.

15 Nerlove, M., 1968, Factors affecting differences among rates of return on investments in individual common stocks, *The Review of Economics and Statistics*, 50, 312–31.

16 Fabozzi, F. and Francis, J., 1979, Industry effects and the determinants of beta. *The Quarterly Review of Economics and Business*, 19, 61–74.

17 Capaul, C., 1999, Asset pricing anomalies in global industry indexes, *Financial Analysts Journal*, July/August, 55, 4, 17–37.

18 Weiss, R.A., 1998, Global sector rotation: new look at an old idea, *Financial Analysts Journal*, May/June, 54, 3, 6–8.

19 One of the sectors included in the analysis involves drilling. As an anonymous referee pointed out, this is a shipping-related business – affecting indirectly the freight market. The demand and supply side of the sector are affected by conditions which are outside those of the freight market. However, the outcome of the sector does affect the supply side of the freight market. Strictly speaking,

this is the case for all sub-sectors distinguished in Table 4.1. For instance, the economics, including the risk-return profiles, for Bulk, Container and Tanker shipping, just to pick three sectors analysed in the paper, are very different between them – see e.g. Kavussanos, M.G., 2002, Business risk measurement and management in the cargo carrying sector of the shipping industry, *Maritime Economics and Business*, edited by C. Grammenos (London: LLP), Ch 30, 661–92. Also, the same argument holds for the Cruise sector, which is related to the economics of the international leisure industry, as well as shipping. However, all these sectors have been selected for analysis, because they all have some impact on various parts – sectors of the shipping industry.

20 *Lloyds List* obtain their stock price information from Bloomberg's classification list.

21 Accordingly, some companies known to have major shipping or shipping-related interests were excluded because they were too diversified elsewhere.

# 5  Perceptions of foreign exchange rate risk in the shipping industry

*K. Akatsuka and Heather Leggate*

## 1 Introduction

The international shipping industry has certain characteristics, which make it particularly vulnerable to exchange rate risk. Whilst the commercial world recognizes such risk, neither the shipping industry nor academia have yet tried to quantify or evaluate the extent of the problem.[1] This chapter attempts to assess this risk from a market perspective by analysing the impact of exchange rate changes on share price returns of major shipping companies from two major shipping nations, Japan and Norway. Although these nations account for 13 per cent and 10 per cent of the world fleet (measured in dead weight tons), respectively, the primary reason for selecting these particular countries is the different patterns of exchange rate fluctuations that have occurred over recent years. The yen has been subject to long periods of appreciation against the US dollar, whilst the krone has exhibited volatility, but no apparent trend.

The measures of performance is returns on shares, a measure which encompasses all activities of a business, trading, sale and purchase, and, more importantly, investor perception. The approach on returns data is, of necessity, company specific, since no meaningful index exists for shipping company shares. However, since the bulk of the industry in the two countries is represented by a few major companies, analysis of these companies will give a good approximation for the industry as a whole.

Whilst the main hypothesis is an analysis of the impact of exchange rate risk on performance, the chapter provides a comparative analysis which allows further hypotheses to be explored. It measures significant differences in the results of the two countries chosen, and, within those countries, significant differences between the companies themselves. Differences between countries may be the result of economic factors and national responses to the exchange rate risk. Variations between the companies may be the consequence of differences in corporate policy, management, and type of trade, for example, liner, dry bulk, tank. Sector specific differences are more difficult to assess since all the companies selected for this analysis have a diverse operational base covering a range of the shipping sectors.[2] In each case, there is a bias towards a particular trade, and this factor is used to analyse such differences.

The chapter begins with a discussion of the origins of exchange rate in section 2, followed by a detailed methodology in section 3. Results are presented in section 4 and conclusions drawn in section 5.

## 2 Exchange rate risk

Foreign risk in the shipping industry arises from a unique freight market structure exposed to a volatile foreign exchange market. The exposure arises from revenues in US dollars, which are not matched by US dollar expenses. The net dollar revenues, i.e. US dollar revenues less US dollar costs, must, therefore, be converted into other currencies to meet those costs not denominated in US dollars, the majority of which will be in the domestic currency. In a system of volatile exchange rates, fluctuations in the rate of exchange between these currencies and the US dollar can, therefore, have a serious impact on the performance of the shipping industry in terms of operating results and returns on shares.

The problem thus consists of two essential elements. First, the exposure in terms of net US dollar revenues, and, second, the volatility of the exchange rate. This study uses a number of primary sources, detailed discussions with shipowners, shipbrokers, and financiers, and official statistics to identify the exposure for the Japanese and Norwegian industries. Both these elements are discussed in what follows.

### 2.1 Exchange rate volatility

The collapse of the fixed exchange rate regime in the 1970s led to instability of currencies on the international market. This study focuses on the events and trends of the last decade, in order to determine the more recent effects of this volatility on the shipping industry.

Figure 5.1 illustrates the volatility of the nominal yen/US$ and krone/US$ exchange rate since 1990. The diagram shows the percentage change in the nominal exchange rates against the US dollar since 1990, with January 1990 as the base year. It can be seen that both the yen and krone exhibit volatility against the US dollar during that period. The yen, however, shows a marked long-term appreciation up to mid-1995, being particularly strong between 1993–95. During 1996, there is a revision back to 1992 levels. The appreciation of 1997 leads up to the Asian Crisis, in which all the Asian currencies depreciated dramatically on the international markets in the wake of a major banking collapse in the Far East. The recovery from this has been relatively strong, with the yen in 1999 reaching its highest levels against the US dollar. The underlying strength of the yen is clearly apparent during the period under investigation. The experience of the krone is less clear, with no apparent trend emerging.

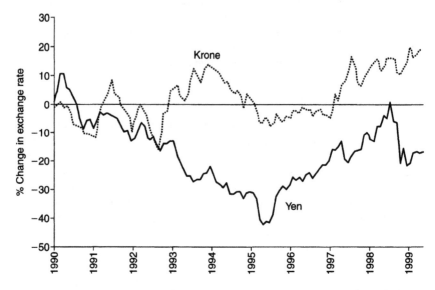

*Figure 5.1* Percentage change in nominal exchange rates against the US dollar 1990–99 (base January 1990) (source: Datastream).

In the early part of the decade, from 1990–93, the fluctuations both upward and downward are around a relatively consistent average rate for the period; 1993 proved to be a year of depreciation against the US dollar, which was reversed in 1994; 1995 and 1996 were characterized by low levels of volatility.

It is not, however, this volatility in itself, but the fact that the companies maintain an exposure to such volatility, which creates the risk.

### 2.2  Exposed flows

The exposed flows in the cases of both Japan and Norway refer to the US dollar denominated operating revenue, which is not used to cover US dollar denominated costs. As far as possible, the industry will engage in a process of natural hedging by using its dollar receipts to meet dollar payments. The remaining net dollar receipts will be converted into other currencies. These are the exposed flows. Clearly, a number of currencies are involved, not just the domestic currency. The percentages for these other currencies are difficult to quantify, and it is assumed, therefore, that all are converted to the domestic currency. This may result in an over- or under-estimate, depending on the volatility of these other currencies compared to the domestic currency. The amounts involved are not expected to be significant, since most of the sums involved are converted to the domestic currency.

The exposed portion of revenues may be determined either by the use of existing aggregate data, or by inference based on detailed discussion with companies and industry specialists.[3] The results of this analysis for the operational side of the business are presented in Table 5.1.

This exposure is calculated as the level of net US dollar revenues,[4] since these are the amounts which must be converted into other currencies. Table 5.1 compares the levels of exposure for both Japan and Norway, as measured by the net US dollar revenues as a percentage of total operating revenues.

The Japanese industry has employed policies to reduce the level of exposure to exchange rate risk by shifting costs into US dollars. With the exception of the 16 per cent level in 1993, the figures in Table 5.1 illustrate the effectiveness of this strategy. The Norwegian industry, on the other hand, maintains a higher level of exposure. Although aggregate data is only available for 1994 and 1995, discussion with Norwegian industry specialists[2] suggests that these levels have been maintained throughout the ten-year period at ~50 per cent.

## 3 Methodology

Given the exposure of the industry to exchange rate volatility, it is interesting to test the hypothesis that such fluctuations have an impact on the performance of the industry as perceived by the market. This involves an examination of the relationship between exchange rate movements and share price returns.

The analysis uses a regression technique whereby the returns on the share are regressed against the changes in the log exchange rate for the respective currency. The regression equation may be formally expressed as follows.[5]

$$R_{i,t} = \alpha_0 + \sum_{j=0}^{n} c_j \Delta ER_{i,t-j} + \epsilon_{i,t} \tag{1}$$

*Table 5.1* Exposure for Japan and Norway

| Year | Net dollar revenues as % of dollar operating revenue | |
|------|-------|--------|
|      | *Japan* | *Norway* |
| 1987 | 22.4 | NA |
| 1988 | 18.4 | NA |
| 1989 | 14.3 | NA |
| 1990 | 10.0 | NA |
| 1991 | 14.2 | NA |
| 1992 | 12.4 | NA |
| 1993 | 16.0 | NA |
| 1994 | 13.3 | 52.7 |
| 1995 | 12.7 | 51.6 |
| 1996 | 14.6 | NA |

where: $R_{i,t}$ is the return for investment during period $t$; $\Delta ER_{i,t-j}$ represents the change in the real US dollar exchange rate index for the period $t$ to $j$; $\alpha_0$, $\beta$ are the parameters to be estimated; and $\epsilon_{i,t}$ is the error term for investment $i$ in period $t$.

The statistics used to measure the relationship are $R^2$, the coefficient of determination, and the $t$ statistic.

This study will examine the relationship between the contemporaneous and lagged change in the value of the US dollar in terms of krone and in terms of yen and the share price returns of the major Norwegian and Japanese shipping companies, quoted on the relevant stock exchanges. The companies analysed are Nippon Yusen Kabushiki Kaisha (NYK), Kawasaki Kisen Kaisha (K Line) and Mitsui OSK Line (MOL) for Japan and Bergesen, Leif Hoegh and Wilhelmsen for Norway. The data is monthly data on share prices and exchange rates against the US dollar obtained from Datastream.[6]

The returns on an investment are taken to be the return from capital gain. Dividends are excluded because they are particularly low for all companies over the period 1987–99. This low level of dividend is a particular characteristic of the Industry. The returns from capital gain may be expressed as:

$$r_{i,t} = \frac{p_{i,t} - P_{i,t-1}}{p_{i,t-1}} \tag{2}$$

where $p_{i,t}$ is the price of the investment at time $t$; and $p_{i,t-1}$ is the price of the investment at time $t-1$.

Similarly:

$$r_{m,t} = \frac{p_{m,t} - p_{m,t-1}}{p_{m,t-1}} \tag{3}$$

where $p_{m,t}$ is the price of the investment at time $t$; and $p_{m,t-1}$ is the price of the investment at time $t-1$.

All data was tested for stationary using a Dickey-Fuller test.[7] All returns data and the first difference of log exchange rate data were found to be stationary.

## 4 Cross-sectional analysis

The following sections examine the impact of log exchange rate movements on share price returns using monthly data over a 164-month period from September 1986 to May 2000. The analysis uses both cross-sectional and time series approaches. This enables examination of the impact in general terms and as a comparative on a company specific and industry specific level.

### 4.1 Company specific comparison

The first set of results examine the separate industries of Japan and Norway in turn, in order to assess the impact of exchange rate changes on the share price returns. Formally, this can be written:

$$R_{ct} = \alpha + \beta ER_{ct} + \epsilon_{ct} \tag{4}$$

for $c$ = company 1, 2 and 3 and $t$ = 1, 2, ... $T$ time periods.

### 4.2 Japan

Since the merger of MOL and Showa and NYK with Navix there are only three 'big' companies in Japanese shipping. This analysis is therefore based on data for the 'big 3', MOL, NYK and K Line. The firm-specific effects are investigated using a $F$-statistic to test whether the constant terms are all equal.

Table 5.2 shows the results of the company-specific analysis for the whole data set from 1986–2000.

The results show that 27 per cent of the log appreciation in the yen against the dollar six months ago, 60 per cent of the log appreciation five months ago, 41 per cent of the log depreciation in the rate three months previously, and 34 per cent of the log depreciation two months previously, affect the current levels of returns. The appreciation has a major positive influence on returns. This seems counter-intuitive, as an appreciation is expected to have a negative effect on performance. This idea is discussed more fully later in this section. The $F$-test on fixed effects shows no significant firm-specific effects. The results of the individual companies within the Japanese industry are not detectably different.

In order to investigate the reason for the apparent anomaly, it was decided to assess the impact of the Asian Crisis on the results by dividing data into two sub periods, pre- and post-crisis. The pre-crisis results are shown in Table 5.3.

*Table 5.2* Japan company-specific: whole period

| Variable | Coefficient | t-value |
|---|---|---|
| Constant | 0.001 465 | 0.276 598 |
| $ER_{t-2, t-3}$ | 0.336 830 | 2.411 266 |
| $ER_{t-3, t-4}$ | 0.410 358 | 2.923 291 |
| $ER_{t-5, t-6}$ | −0.604 061 | −4.285 930 |
| $ER_{t-6, t-7}$ | −0.276 293 | −1.986 220 |

Notes
Sample 1986M9 to 2000M5.
Dependent variable: returns on shares.
$R^2$ = 0.068 145.

Table 5.3 Japan company-specific: pre-Asian crisis

| Variable | Coefficient | t-value |
|---|---|---|
| Constant | 0.002670 | 0.453811 |
| $ER_{t-3,t-4}$ | 0.579903 | 3.468248 |
| $ER_{t-6,t-7}$ | −0.318047 | −1.917892 |

Notes
Sample 1986M9 to 1997M5.
Dependent variable: returns on shares.
$R^2 = 0.40911$.

In this case, the significant lags are at six months, with an appreciation in the yen/US$ rate, and at three months, with a larger depreciation having an impact on current returns. The dominant effect for this period is the depreciation in the yen. Again, there are no firm-specific effects as evidenced by the $F$-test.

## 4.3 Post-Asian crisis

An examination of the post-crisis period reveals a more rapid response to exchange rate movements, since it is the changes of five months ago and two months ago which have an effect on the current level of returns. Again, there is evidence that an appreciation of the yen five months previously and a smaller depreciating movement two months ago have a positive impact. In this period, however, the overall effect is one of appreciation affecting the current level of return, as illustrated in Table 5.4.

An examination of the effects of the different companies reveal no significant differences. The comparative analysis of the Japanese situation indicates no differences in the way in which exchange rate movements affect the returns for MOL, NYK and K Line. However, the Asian crisis does appear to influence the results. In the pre-crisis period, the depreciation of the yen has the dominant effect, whereas the post-crisis period sees the positive effect of an appreciation on returns. Furthermore, the impact is more immediate following the crisis. These differences between the pre-

Table 5.4 Japan company-specific: post-Asian crisis

| Variable | Coefficient | t-value |
|---|---|---|
| Constant | 0.001301 | 0.116930 |
| $ER_{t-2,t-3}$ | 0.702429 | 2.878577 |
| $ER_{t-5,t-6}$ | −1.092504 | −4.445744 |

Notes
Sample 1997M6 to 2000M5.
Dependent variable: returns on shares.
$R^2 = 0.242551$.

and post-crisis results are investigated using a Wald test on the coefficients. This test indicates significant differences between the coefficients of the variables, and is carried out by isolating the pre- and post-crisis exchange rate movements. Where a particular lag is significant in either case, be it in the pre- or post-crisis period, it is included in the regression. Table 5.5 shows the results of this analysis. The dummy variable is set to 0 in the pre-crisis period and 1 in the post-crisis period.

The results of the test indicate the significance of the three and six months lags for the pre-crisis period and the two and five month lags for the post-crisis period. The Wald test below is based on a null hypothesis that the coefficients pre- and post-crisis are the same. This hypothesis is clearly rejected and, thus, there is a significant difference between the results of the two periods.

### 4.4 Norway

The same analysis is performed using the data for the Norwegian industry, first the whole period and then an analysis between the pre- and post-crisis periods. The results of the regression for the whole period from 1997–2000 are shown in Table 5.6.

The results show a strong and significant contemporaneous and two-month lagged effect of the exchange rate movement on the returns.

*Table 5.5* Wald test on coefficients pre- and post-crisis

| | |
|---|---|
| Null hypothesis: | |
| $preER_{t-2,t-3} = postER_{t-2,t-3}$ | |
| $preER_{t-3,t-4} = postER_{t-3,t-4}$ | |
| $preER_{t-5,t-6} = postER_{t-5,t-6}$ | |
| $preER_{t-6,t-7} = postER_{t-6,t-7}$ | |
| $F(4,491)$ calculated | 5.136150 |
| $F(4,491)$ critical 5% | 2.37 |
| Reject null hypothesis | |

*Table 5.6* Norway company-specific: whole period

| Variable | Coefficient | t-value |
|---|---|---|
| Constant | 0.007876 | 1.634575 |
| $ER_{t,t-1}$ | 0.974289 | 5.717848 |
| $ER_{t-2,t-3}$ | 0.503364 | 2.913190 |
| $ER_{t-6,t-7}$ | 0.366967 | 2.135900 |

Notes
Sample 1987M8 to 2000M4.
Dependent variable: returns on shares.
$R^2 = 0.092402$.

The difference in results emerging from the cross-sectional analysis compared to the individual company regressions is the significance of the six-month lag, illustrating that some information regarding exchange rate changes take a longer time to affect the share price. As is the case with Japanese industry, there are no significant differences between the results of the companies operating within the industry, as evidenced by the $F$-test.

For consistency, the data is analysed between the same sub periods as in the Japanese case, in order to detect any differences between the variables obtained and different company-specific effects in those periods. Table 5.7 shows the results for the pre-crisis period.

These results are consistent with those for the whole period analysis. The post-crisis results are shown in Table 5.8.

Although there are again no firm-specific differences in the post-crisis period, the results are somewhat surprising, since the Asian crisis was not expected to have an impact on the experience of the Norwegian industry. The outcome suggests that the impact is slower for Norway in this period since there is no contemporaneous link between exchange rate movements and share prices, but a relationship, which is lagged by four months. The difference between the results pre- and post-crisis was again tested using a Wald test. The results are shown in Table 5.9

Table 5.9 confirms the significant differences between the variables pre- and post- the Asian crisis. In the pre-crisis period, the impact is immediate, with the current movements in exchange rates having the greatest effect.

*Table 5.7* Norway company-specific: pre-crisis period

| Variable | Coefficient | t-value |
| --- | --- | --- |
| Constant | 0.013563 | 2.459676 |
| $ER_{t,t-1}$ | 1.281004 | 6.862243 |
| $ER_{t-2,t-3}$ | 0.506456 | 2.652860 |
| $ER_{t-6,t-7}$ | 0.710351 | 3.681809 |

Notes
Sample 1987M8 to 1997M5.
Dependent variable: returns on shares.
$R^2 = 0.159685$.

*Table 5.8* Norway company-specific: post-crisis period

| Variable | Coefficient | t-value |
| --- | --- | --- |
| Constant | −0.004125 | −0.464061 |
| $ER_{t-4,t-6}$ | 0.734838 | 2.156052 |

Notes
Sample 1997M6 to 2000M4.
Dependent variable: returns on shares.
$R^2 = 0.042012$.

*Table 5.9* Wald test on coefficients pre- and post-crisis

| | |
|---|---|
| Null hypothesis: | |
| $\text{pre}ER_{t,t-1} = \text{post}ER_{t,t-1}$ | |
| $\text{pre}ER_{t-2,t-3} = \text{post}ER_{t-2,t-3}$ | |
| $\text{pre}ER_{t-4,t-5} = \text{post}ER_{t-4,t-5}$ | |
| $\text{pre}ER_{t-6,t-7} = \text{post}ER_{t-6,t-7}$ | |
| $F(4,416)$ calculated | 4.6535 |
| $F(4,416)$ critical 5% | 2.37 |
| Reject null hypothesis | |

After the crisis, however, there is only one significant lag, which shows that the change four months previously affects the current level of returns.

The above cross-sectional analysis confirms that exchange rate movements do affect the share price returns of the industry and the individual companies operating in it. In analysing the two industries, there is no company-specific effect in either case. In the model developed, there are, however, clear differences in the significant lags for the two industries, and indeed in the pre- and post-crisis periods. The significance of these is examined in the following section on industry specific differences.

### 4.5 Industry differences

The industry-specific comparison investigations whether there are significant differences between the results of the two industries. Again, the regression model incorporates time series and cross-sectional data as follows:

$$R_{it} = \alpha + \beta ER_{it} + \epsilon_{it} \tag{5}$$

for $i$ = industry, Japan or Norway and $t = 1, 2, \ldots T$ time periods.

Statistically, the objective can be expressed as follows:

$$
\begin{bmatrix} R_{J1} \\ \cdot \\ \cdot \\ \cdot \\ R_{Jn} \\ R_{N1} \\ \cdot \\ \cdot \\ R_{Nn} \end{bmatrix}
=
\begin{bmatrix} 1 & 0 \\ \cdot & \cdot \\ \cdot & \cdot \\ 1 & 0 \\ 0 & 1 \\ \cdot & \cdot \\ \cdot & \cdot \\ 0 & 1 \end{bmatrix}
\begin{bmatrix} ER_{J1} & 0 \\ \cdot & \cdot \\ \cdot & \cdot \\ ER_{Jn} & 0 \\ 0 & ER_{N1} \\ \cdot & \cdot \\ \cdot & \cdot \\ 0 & ER_{Nn} \end{bmatrix}
\begin{bmatrix} \alpha_J \\ \alpha_N \\ \beta_J \\ \beta_N \end{bmatrix}
\begin{bmatrix} \epsilon_{J1} \\ \cdot \\ \cdot \\ \epsilon_{Jn} \\ \epsilon_{N1} \\ \cdot \\ \cdot \\ \epsilon_{Nn} \end{bmatrix}
\tag{6}
$$

where: $R_J$ is the return on the shares of the Japanese companies for time periods 1 to $n$; $R_N$ is the return on the shares of the Norwegian companies for time periods 1 to $n$; $ER_J$ is the yen/US\$ exchange rate for time periods 1 to $n$; $ER_N$ is the krone/US\$ exchange rate for time periods 1 to $n$; $\alpha_J$ is the constant for Japan; $\alpha_N$ is the constant for Norway; $\beta_J$ are the coefficients for Japan; $\beta_N$ are the coefficients for Norway; $\epsilon_J$ is the error term for Japan; and $\epsilon_N$ is the error term for Norway. The null hypothesis $(H_0)$ is: $\alpha_J = \alpha_N$ and $\beta_J = \beta_N$.

Again, the analysis uses three stages: the first for the whole period, the second and third for the pre- and post-crisis periods. All regressions test for differences between the constant term and the coefficients of the regressors. The constants are tested using a dummy variable, where the dummy is equal to 1 for the Japanese industry and 0 for the Norwegian industry. The coefficients are tested by isolating the lagged variables and performing a Wald test.

The results of the regression for the whole period are examined in Table 5.10.

The significant variables are the current change in the exchange rate, and the changes of two, three and five months previously. The regression again shows the anomaly seen in the Japanese case, where an appreciation in the exchange rate has a positive effect on current returns. The fact that the overall impact is one of depreciation suggests that this be adjusted by the subsequent movements.

It is interesting, however, that the dummy variable is not significant according to the $t$-statistic. In other words, there is no difference between the constant terms. This is confirmed by an $F$-test, which examines the $R^2$ of the unrestricted and restricted models and compares it to the calculated $F$ with the critical $F$-statistic from the appropriate tables.

The significant differences between the coefficients are tested using a Wald test.

Separating all the variables and lags of variables between Japan and Norway facilitates the testing for different coefficients of the two models, as shown in Table 5.11. The results show different significant variables for

*Table 5.10* Industry comparative: whole period

| Variable | Coefficient | t-value |
|---|---|---|
| Constant | 0.008024 | 2.175411 |
| $ER_{t,t-1}$ | 0.266449 | 2.398208 |
| $ER_{t-2,t-3}$ | 0.368658 | 3.352999 |
| $ER_{t-3,t-4}$ | 0.312095 | 2.853263 |
| $ER_{t-5,t-6}$ | −0.348815 | −3.170997 |

Notes
Sample 1986M8 to 2000M4.
Dependent variable: returns on shares.
$R^2 = 0.039513$.

*Table 5.11* Wald test on coefficients for Japan and Norway

| | |
|---|---|
| Null hypothesis: | |
| $YER_{t,t-1} = KER_{t,t-1}$ | |
| $YER_{t-2,t-3} = KER_{t-2,t-3}$ | |
| $YER_{t-3,t-4} = KER_{t-3,t-4}$ | |
| $YER_{t-5,t-6} = KER_{t-5,t-6}$ | |
| $YER_{t-6,t-7} = KER_{t-6,t-7}$ | |
| $F(5,913)$ calculated | 6.451708 |
| $F(5,913)$ critical | 2.12 |
| Reject null hypothesis | |

Japan and Norway, which are consistent with the separate industry analysis presented earlier in the chapter.

The Wald test above starts with the hypothesis that the coefficients of the variables are equal for both countries and shows that there is a significant difference between the variables of the two industries, Japan and Norway, for the whole period from 1987–2000.

### 4.6 Asian crisis

The next section divides the data into two sub periods, to cover the results pre- and post-Asian crisis, since this factor proved to be influential in the results of the Japanese industries. This analysis follows the same procedure as that outlined above, testing first the constants then the coefficients of the variables. Table 5.12 shows the results for the pre-crisis period.

The regression produces different results. Both the current and the three-month lag are significant and, in both cases, it is a depreciation in the domestic exchange rate which has a positive impact on returns. The dummy variable is not significant in the pre-Asian crisis regression and is, thus, not included in the above table. Again, this is confirmed by the *F*-test.

The reorganized data is used to test the coefficients, with results shown

*Table 5.12* Industry comparative: pre-Asian crisis

| Variable | Coefficient | t-value |
|---|---|---|
| $ER_{t,t-1}$ | 0.326802 | 2.510833 |
| $ER_{t-3,t-4}$ | 0.387640 | 2.973076 |
| Constant | 0.012557 | 2.952969 |

Notes
Sample 1986M8 to 1997M5
Dependent variable: returns on shares.
$R^2 = 0.020242$.

in Table 5.13. The results of the test are again consistent with the individual industry regressions. The Wald test on the coefficients show that there is a significant difference between the results of Japan and Norway for the pre-crisis period.

### 4.7 Post-Asian crisis

The period following the Asian crisis inevitably has fewer data points. However, at over 30 observations for each company, this is sufficient to gain a meaningful result. The sample period starts from June 1997.

The post-crisis data shows the positive impact of the five-month lagged appreciation in the exchange rate. There is a part adjustment three months later of a 70 per cent change in the log exchange rate, but the overall effect is negative. The result suggests that in the post-crisis period an appreciation in the exchange rate can be seen as very positive (see Table 5.14). This is not so surprising following the dramatic fall in business confidence during the crisis. Against such a background, an appreciation in the exchange rate is indicative of improving business and economic conditions. As far as the market was concerned, this factor outweighed the potentially damaging impact of the appreciating yen for the maritime industry.

In this period, the same dummy variable is used, but the coefficient is not significantly different from zero and is, therefore, excluded from the final regression.

*Table 5.13* Wald test on coefficients for Japan and Norway

| | |
|---|---|
| Null hypothesis: | |
| $YER_{t,t-1} = KER_{t,t-1}$ | |
| $YER_{t-2,t-3} = KER_{t-2,t-3}$ | |
| $YER_{t-3,t-4} = KER_{t-3,t-4}$ | |
| $YER_{t-6,t-7} = KER_{t-6,t-7}$ | |
| $F(4,704)$ calculated | 11.09973 |
| $F(4,704)$ critical | 2.37 |
| Reject null hypothesis | |

*Table 5.14* Industry comparative: post-Asian crisis

| Variable | Coefficient | t-value |
|---|---|---|
| Constant | 0.001301 | 0.116930 |
| $ER_{t-2,t-3}$ | 0.702429 | 2.878577 |
| $ER_{t-5,t-6}$ | -1.092504 | -4.445744 |

Notes
Sample 1997M6 to 2000M5.
Dependent variable: returns on shares.
$R^2 = 0.242551$.

The *F*-test above demonstrates that there is no significant difference between the constants of Japan and Norway. The Wald test shown in Table 5.15 examines differences between the coefficients. As expected, it shows significant differences between the coefficients in the post-crisis period.

The lags are found to be significantly different for the two industries, Japan and Norway, and indeed for the pre- and post-crisis periods. In the Norwegian case, the current movements together with the two- and six-month lags in the exchange rate have an effect on current returns. The contemporaneous lag is not surprising when considering the high level of exposure, which the Norwegian industry sustains. However, the net profit margins for the Norwegian companies are higher than those of the Japanese, which would lower the expected impact of exchange rate movements as far as shareholders are concerned. In both cases, the Asian crisis has a notable influence on the results, but the impact is more dramatic in the Japanese experience. In the post-crisis period, the effect of an appreciation in the yen and the business confidence that this exudes outweigh the negative impact of appreciation on performance. The effect is slower than the Norwegian case, which is consistent with the lower levels of exposure, but seems surprising when taking into account the very low net profit margins experienced by the Japanese companies (see Table 5.16). With such low margins, a small change in the exchange rate can have a huge impact on performance in terms of profits available to shareholders.

### 4.8 Sector analysis

A further hypothesis is that the type of trade, namely: liner, dry bulk, tanker, car carrier, has an impact on the results. This is more difficult to test, since all the companies are diversified. In other words, their activities are not limited to any one sector. In some cases, there is even diversification into non-maritime activities, such as logistics and freight forwarding, although these trades are closely related to the maritime industry. Table 5.17 presents an analysis of the dominant trade for each company based on either operating revenue or operating profit.

*Table 5.15* Wald test on coefficients for Japan and Norway

| | |
|---|---|
| Null hypothesis: | |
| $YER_{t-2,t-3} = KER_{t-2,t-3}$ | |
| $YER_{t-4,t-5} = KER_{t-4,t-5}$ | |
| $YER_{t-5,t-6} = KER_{t-5,t-6}$ | |
| $F(3,208)$ calculated | 5.061 411 |
| $F(3,208)$ critical 5% | 2.60 |
| Reject null hypothesis | |

*Table 5.16* Net profit margins (profit after tax)

|      | MOL   | NYK   | KLine | BERG  | LEIF  | WILH  |
|------|-------|-------|-------|-------|-------|-------|
| 1987 | −1.66 | −0.57 | −2.15 | 9.68  | NA    | −4.4  |
| 1988 | −2.77 | 0.28  | −2.56 | 2.70  | NA    | −0.9  |
| 1989 | 1.02  | 0.55  | −0.15 | 3.77  | NA    | 2.54  |
| 1990 | 2.87  | 1.52  | 0.64  | 17.21 | −4.05 | 2.42  |
| 1991 | 0.95  | 0.78  | −0.53 | 3.85  | 6.15  | 2.62  |
| 1992 | 0.47  | 0.78  | 0.97  | 0.95  | 8.29  | 4.37  |
| 1993 | 1.12  | 0.85  | −0.46 | 4.55  | 10.03 | −2.29 |
| 1994 | −0.95 | 0.70  | 0.60  | 3.11  | 8.68  | 4.97  |
| 1995 | −0.68 | 0.45  | 0.80  | 8.56  | 10.36 | 3.24  |
| 1996 | 0.67  | 0.33  | 0.64  | 35.54 | 21.49 | 2.42  |
| 1997 | 0.77  | 1.37  | 1.37  | 11.41 | 9.64  | 13.27 |
| 1998 | 1.22  | 0.73  | 0.37  | 17.44 | 10.63 | 6.99  |
| 1999 | 0.85  | 1.14  | 0.38  | 7.65  | NA    | NA    |

Source: Annual reports.

*Table 5.17* Types of trade

| Company | Dominant trade | Percentage |
|---------|----------------|------------|
| MOL     | Liner          | 41.3       |
| NYK     | Bulk and car carrier | 57.6 |
| K Line  | Liner          | 46.8       |
| BERG    | Tankers        | 50.3       |
| LEIF    | Car carriers   | 57.3       |
| WILH    | Liners         | 94.3       |

Source: Annual reports 1999 for Japanese companies, 1998 for Norwegian companies.

The hypothesis that the sector has an influence was tested by setting up a dummy variable for the liner sector, and by running the regressions for all the data. In all cases, the dummy variable was found to be not significantly different from zero, suggesting that type of trade does not affect the result. In this case, the result is an inference, because of the fact that the analysis is based on corporate data and there is diversification in all cases.

## 5 Conclusion

This chapter has analysed the market perception of exchange rate risk in the shipping industry by examining the relationship between these exchange rate movements and share price returns and finds this risk to be significant in the determination of performance. It has thus provided a comparative analysis of the impact of exchange rate movements on returns of two industries, Japan and Norway, and the major companies operating in them. It finds that there are significant differences between the experience of the two industries, Japan and Norway, but that there are no significant company-specific effects within those industries. Although all the companies operate across a diversity of trades within the industry, it is possible to draw some conclusion about the impact of sectors on the analysis from the dominant trade of each. This comparison also shows no significant differences between the experience of the different sectors.

The difference between the pre- and post-Asian crisis experiences is interesting in both cases. In the Norwegian case, it is surprising to find that the crisis has any impact at all, but the results strongly indicate that the reaction to exchange rate movement is much slower after 1997. The effect is more dramatic in the Japanese case, where an appreciation of the yen yields positive returns and a more immediate effect. In the wake of the crisis, it appears that the business confidence exuded by such an appreciation outweighs the costs for the maritime industry.

Measurement of the impact of exchange rate volatility on the performance of these two major shipping companies reveals its remarkable magnitude. The exposure appears to play a very important role. For the Norwegian industry, where exposure is high at ~50 per cent, the impact on returns is felt more immediately, with a substantial contemporaneous depreciation of the krone affecting the current returns. For Japan, where the exposure levels are much lower, the impact on share price returns is less marked. In fact, there is evidence that an appreciation has a positive impact. This apparently perverse situation can be explained partly by the low exposure, and partly in terms of improving business and economic conditions, particularly in the wake of the Asian crisis. In the post-crisis period, the confidence exuded by an appreciating yen has the greater influence on returns even for the maritime industry.

The results indicate the importance of the strategic response to exchange risk, the consequences of the active Japanese policy of shifting as many costs as possible into US dollars, in an attempt to minimize exposure, and the effects of the Norwegian companies' decision to maintain their exposure, in order to profit by speculation on the movement in the rate of exchange. Such strategies have been largely formulated in the light of recent history. During the 1980s and 1990s, the Japanese shipping industry was prey to the effects of long-term appreciation of the yen, whereas no such trend emerged for the krone.

The analysis of the fundamental and dramatic effects of foreign exchange risk on the performance of two major shipping nations was used here to exemplify the effects of such risks on world shipping in general. The degree of impact is clearly related to level of exposure. Exposure itself can not always be seen as a negative, since it allows the industry to take advantage of favourable movements. High levels of exposure allow greater benefit from these favourable movements, but require active management to minimize the downside risk. Lower exposure, on the other hand, allows a more passive approach to be taken. Whatever its level, some exposure is inevitable, and requires effective management. In an industry, which is particularly vulnerable to the vagaries of the market for foreign exchange, the shipping industry must take this management extremely seriously.

## References and notes

1 Beth, H.L., 1979, Fluctuations of exchange rates – their impact on shipping, Institute of Shipping Economics, Bremen, Lectures and Contributions No. 26, was one of the first academics to discuss the problem in relation to the shipping industry.
2 See Table 5.17.
3 These specialists include company directors, treasurers, and financiers.
4 This is calculated as the dollar operating revenues less the dollar operating expenses.
5 Bartov, E. and Bodnar, G.M., 1994, Firm valuation, earnings, expectations, and exchange rate exposure effect, *Journal of Finance*, 49, 1755–85 used a similar methodology but used abnormal returns on the basis that the market does not price the risk. Research has shown that this is not the case with these companies in the Shipping Industry (Leggate, H.K., 2000, The Impact of Exchange Rate Risk on the Shipping Industry, PhD Thesis, London Guildhall University).
6 Datastream: Codes: Yen/US$ exchange rate YAPAYE$, Krone/US$ exchange rate MBNOKSP, MOL J:MON@N 932300, NYK J:NY@N 930449, K Line J:KK@N 932298, Bergesen N:BEA 745249, Leif Hoegh N:LHO 775919, Wilhelmsen N:WWI 929581.
7 Dickey, D.A. and Fuller, W.A., 1979, Distribution of the estimators for autoregressive time series with a unit root, *Journal of the American Statistical Association*, 74, 427–31; Dickey, D. and Fuller, W., 1981, Likelihood ratio statistics for autoregressive time series with a unit root, *Econometrica*, 49, 1057–72.

# Part II

# People and skills

*Edited by Valentina Carbone*

# 6 Developments in the labour market

*Valentina Carbone*

## 1 Introduction

In shipping, the factors of production are mainly the vessel (capital), the crew (labour). No matter how simple or complex the physical marine asset, the true, primary assets of the company are the people who crew the vessel and who are in charge of the navigation and operation of the vessel. Without them, and without their operating at their highest level of functional ability, the company's physical assets, profits and commercial survival are extremely vulnerable.

International shipping has been undergoing significant changes in the past three decades, affecting the way to optimise the interaction of the above-mentioned factors of production. Globalisation is generally considered to be fundamental in this system of change.

Before the 1970s, the nations that built the ships, usually owned the ships, registered the ships, serviced the ships, crewed the ships, trained the crews, supervised the performance of crew and ships, and often provided the cargoes inbound or outbound.

Then came the oversupply of shipping that followed the oil crisis of the 1970s and the desperate drive to cut costs. This, coupled to the increasing social costs in the mid-1970s, led owners to register their ships offshore so that crew costs could be cut by enabling owners to use foreign crews rather than their own higher paid nationals.

Article 91 of the Law of the Sea Convention states that:

> every state shall fix the conditions for the grant of its nationality to ships, for the registration of ships in its territory, and for the right to fly its flag. Ships have the nationality of the State whose flag they are entitled to fly. There must exist a genuine link between the State and the ship.

However the definition of 'a genuine link' has never been defined.

The growth of offshore flags provided an initial reduction in costs but severed the common link of citizenship between seafarers, trainers, employers, administrators, supervisors, ship owners and often cargo owners. The rapid expansion in the use of crews from developing nations, especially ratings resulted in lower wage costs, often little training, minuscule monitoring of crew treatment and new flag states that exercised token supervision of the ships they registered.

These changes created a global market for the supply of seafarers with the major determinant being the price of labour. By now at least two thirds of all seafarers in the world fleet are employed on ships with multinational crews. Nevertheless, the global market for seafarers is far from homogeneous where legislation, access to work and living conditions are concerned.

In the following section are presented some of the contributions dealing with the labour market and trying to quantify the demand and the supply of seafarers; then we will analyse some of the main issues for manpower in shipping, e.g. deteriorating quality of working and living conditions and the increasing importance of safety onboard. Finally, Maritime Education and Training and Culture are identified as some possible fields of actions to improve conditions for manpower in the shipping industry.

## 2 Demand and supply of seafarers

Historically, the market has been dominated by the traditional developed shipping countries (TDSCs) throughout the post-Second World War era. Although the national flag fleets of the TDSCs have been declining in terms of size, and the number of vessels registered thereto, in terms of actually *controlled* fleet, the TDSCs still maintains a dominant position in the shipping marketplace.

The undergoing structural modification may be partly attributed to the emergence of developing countries (DCs) and the newly-industrialised countries (NICs) as new maritime powers, and to the attempts of companies in industrialised countries to reduce operative costs through the employment of cost competitive foreign seafarers, which has given rise to the above-mentioned phenomenon of *flagging out*. Some studies indicate that cost factors for *flag out* decisions relate more to operational expenses (vessel maintenance), less to crewing and fiscal costs.

The exuberance of Flag of Convenience (FOC) has polarised the countries into seafarer demand countries and seafarer supply countries. The BIMCO/ISF study has documented the ongoing decline of seafarers from the well-developed, traditional maritime nations,[1] while the economic and social benefits of pursuing a seafaring career are readily observed in major supplier nations such as the Philippines and India. The role of non-traditional seafarers supply sources will continue to increase because of the worsening supply shortage of navigating officers and marine engineers.

The long-term trends of employment in the industry have demonstrated a continuing movement towards the utilisation of crews from labour supplying countries. Recently, there is evidence of heightened competition between these countries with regard to seafarers' employment opportunities, notably between China and the Philippines.[2]

The BIMCO Manpower Update of Year 2000 identified a worldwide shortage of 16,000 competent engineers and officers, expected to reach 46,000 by the year 2010. The same study estimated the world supply of seafarers in Year 2000 at 1,227,000 comprising 404,000 officers and 823,000 ratings. The global supply of ratings (823,000) exceeded the demand by 264,000. The general conclusion has been that there is a growing shortage of competent engineers and officers with a heavy and expanding oversupply of ratings.

Despite the unique value of BIMCO/ISF survey as a source for further maritime labour research, some authors reach diverging estimations and/or suggest improvements in order to obtain a more reliable picture of the labour market. Even the authors of the BIMCO Update made it clear that there was a paucity of precise reliable data on the numbers of seafarers employed or available from individual labour supply nations.

Among those who report different figures from BIMCO, it is worth mentioning Leggate (2004)[3] who assesses the demand and supply of seafarers both at a global scale, and at country level. Such contribution is based on the ILO research commissioned in 2001 which challenges the results of the BIMCO/ISF manpower update, as the total number of seafarers for 35 common flag states is 119 per cent higher.

Among those who provide a methodological insight, Li and Wonham (1999)[4] propose the concept of 'active seafarer' instead of 'qualified seafarer' as a more accurate concept for studies on maritime labour supply/demand. They first stated that the main issue is not the quantity, but the quality of the crew. It seems also that the BIMCO method of treatment of missing data and of Chinese records led to an underestimation in the world supply of manpower and to an overestimation of the future demand.

Li and Wonham argue that the future demand for seafarers does not correlate with fleet growth because of the constant decline in average manning levels due to the increasing use of technology, but this is a debatable issue. As a consequence, in terms of maritime policy a balance should be struck between promoting fleet growth and creating maritime labour market problems on the one hand, and consolidating the fleet while satisfying the demand for officers on the other.

A study on the Greek fleet[5] identifies the decline of the Greek registered and owned merchant fleet as one of the main factors affecting the demand for sea labour. A second factor is thought to be the increase in the size of ships, which is linked to a third factor that is the higher use of modern cargo handling and propulsion systems. Such modern ships coupled with technological and organisational changes often require smaller, however better trained, crews than are needed on existing smaller ships. Shipping becomes more competitive but at the expenses of those employed as seamen.

Other studies have dealt with the issue of the quantity of active seafarers, both on a global scale and at country level. Lin Wang and Chiang (2001) discuss the deck officer structure, demand and supply in Taiwan, where the demand is, on average, 15 per cent higher than supply, every year, due to expanded economic development in the late 1900s and the subsequent increase in the number of vessels.[6]

All in all there seems to be a stringent need for clear and reliable figures and statistics on seafarers. Nevertheless, beyond all manner of estimations proposed by those interested in the subject, there seems to be an agreed view on the enhancing industry's dependence on the global seafarer in future years. The risk of such a system is the deterioration in quality of seafaring human resource, which decides the quality level of the shipping services. This is a real concern in the industry, also due to the contemporaneous need for highly educated and land-based personnel to cope with technological and organisational changes in the structure of the shipping industry.

A broader 'qualitative' interpretation of the shortage of quality-educated officers can be found in 'A system dynamic analysis of officer manpower in the merchant marine' (Obando-Rojas *et al.*, 1999)[7] which provides the reader with a systemic interpretation of the flow of people and the flow of information through the manpower supply chain. The analysis, conducted through a rigorous and multidisciplinary approach based on the system dynamic methodology, is applied to the officer supply-chain in the British merchant marine.

Still within such systemic view of shipping, the non-transferability of skills outside the industry makes the shortage of qualified manpower more severe. A fundamental difference between shore industries and shipping is that, in the industry ashore, wastage is a two-way flow (men leaving may be replaced by men trained in other industries), while in shipping it is a one way flow, i.e. when trained officers and ratings leave the sea they cannot be replaced by men trained in other industries.[8] Seafarer recruitment in shore-based jobs makes the number of available trained officers poorer. With regards to the British market, actions to avoid the loss of the current skill base have been proposed.[9] Some of the actions refer to the implementation of a fast-track training programme for graduate entrants, and to the development of training programmes for non-seafarers to fill some of the posts currently filled by former ships' officers when they become vacant.

Contracting sea service life of seafarers and differences in the average age of officers between OECD countries and Far-East and India add to the complexity of the global market for seafarers. The BIMCO update reveals high drop out rates both during Maritime Education and Training (MET) (30 per cent) and subsequently.

Leggate and McConville (2002)[10] reported that 76 per cent of Filipino seafarers are between 25 and 44 years of age. Little opportunity exists for Filipinos beyond 45 years of age. They also reported that the average Chinese seafarer works at sea for from five to eight years.

The BIMCO/ISF study indicates that the average age of officers from OECD countries serving worldwide is considerably higher than officers from the Far East and the Indian subcontinent, showing that the relative lack of supply of European seafarers will accelerate in the future as older seafarers retire. It is argued that additional research is necessary under existing EU programmes to examine the reasons and remedies for this situation so that action can be taken at national and, where appropriate, at EU level to remedy this problem.

In order to look into the contracting sea service life, Glen (1999)[11] tested the statistical significance of a number of characteristics that may help shape the employment pattern of an individual seafaring officer, such as flag, organisational and ship types. Ship type appeared to display the highest effect. This characteristic can be seen as a proxy for different seafaring labour markets, which may require different employment patterns to serve them efficiently.

Some experts propose that a way to reduce the problem of the shortage of skilled seafarers could be found in better recruitment and retention of women at sea. On average, women only account for about 7.6 per cent of the total seafaring labour force in EU ships.[12] Yet the route to cultural change is impeded at all turns. On the Indian subcontinent, for example, many MET institutions are not allowed to recruit women to nautical courses.

Finally, we should mention that, in the past it has been claimed that seafaring was not to be intended as a lifelong occupation but as an interlude in the occupational life of the majority of European seafarers. The decision to pursue employment strategies inspired by such view of the profession, and to accept a high turnover could be very risky, as the safe operation of a ship not only requires trained but also experienced personnel.

## 3 Working and living conditions

Until 30 years ago, in developed nations going to sea to become an officer and captain was seen as an attractive career, an opportunity to see the world, to advance later to a shore-based occupation. For officers and engineers, it was an occupation of respect and dignity that paid better than available shore jobs. In developed nations that is no longer the case.

Today with quick turnaround times, remote terminals, and lesser competence levels among crew, added responsibilities and ship managers driven by desire to cut costs, going to sea has lost its appeal.

Moreover, the nature of the work is such that seafarers have to migrate from one place to another. Seafarers are unquestionably unique migrants (or emigrants) who enter and participate in the country of destinations' workforce (labour market) without entering the country. This is accomplished by working on a vessel whose nationality is indicated by the register and flag and which differs from that of the seafarer's home country.

The mobility of ships and seafarers and the decreasing incidence of intervention by national states and para-national organisations allow shipowners some freedom to set conditions of employment. These are therefore highly variable according to, *inter alia*, country, type of ship, trade route, etc. Consequently, labour markets remain profoundly segregated by national and geographical borders.

Even in the unique international seafaring labour market, there are a number of legislative issues which segment the seafaring labour force. These involve restrictions on the use of foreign seafarers (for example five flags require an all national crew: USA, Italy, India, Mexico and Pakistan),[13] differing taxation systems and differing minimum ages for going to sea. Other indirect barriers may be absence of legislation covering non nationals or changing terms and conditions offered to the different groups.

The most widespread form of discrimination however, is shown in terms of wage differentials between different nationalities. Where there is a relatively large pool of adequately skilled and mobile labour, that is the case of the shipping industry, the wage and employment market is more likely to be highly competitive, if not volatile.

Other social and cultural barriers also exist such as communication skills and social structures. Such frictions discourage the free movement of labour and the employment of any nationality of seafarers on any flag register and serve to create market segmentation.[14]

In such an environment, trade unions could counteract shipowners' power and increase the wage level above what it would otherwise be. Nevertheless, trade unions are mostly national organisations trying to fight against international companies and organisations. International representation is needed to provide effective protection in an international environment. To this end there is a heavy reliance upon the International Transport Workers Federation (ITF) agreements and international conventions drafted by the International Labour Organisation (ILO).

Again, cost reduction seems to be the crucial factor affecting working conditions for seafarers. An OECD study (2001)[15] has confirmed that there is a direct economic incentive in running a substandard ship. The conclusion was that the costs to the users of substandard ships were minimal because the insurance for the ship picked up the bill if anything went wrong. The insurance industry spreads its losses and so it is the wider community that bears the financial costs of substandard shipping, while it is the crews who bear the human costs.

In addition, free choice of flag state as well as the introduction of international registers in traditional seafaring states made it possible to react to the hardening of competition by doing away with many improvements which, since the end of the 1950s, had been introduced aboard merchant ships of highly industrialised countries.

Ratification and acceptance of norms still remains a concern. There are some 30 ILO Conventions and more than 20 recommendations that address sea transport. Of these the ILO 147 Convention of 1976 governing employment and accommodation conditions on board ship is the principal Convention. At July 2002, a total of 43 nations covering only 50.25 per cent of the world fleet had ratified the Convention.

A Protocol to the Convention was adopted in 1996. Together they relate to minimum age, medical examinations, repatriation, articles of agreement, food and catering, accommodation, recruitment and place-ment, sickness and injury, prevention of occupational accidents, welfare, social security, and working hours.

It is worth observing that the world's largest registry, Panama, has not ratified ILO and Japan, the nation that owns 40 per cent of the tonnage under the Panamanian flag does not inspect for ILO147 compliance. Besides, if ratified by each nation, the International Safety Management Code (ISM) would effectively address almost all human problems associ-ated with the operation of ships.

Against such background, the International Commission on Shipping (INCOS) was initiated by the International Transport Workers Federation in 1999 to investigate and appraise the current approach used by governments, industry and interested parties to achieve compliance with international minimum safety, environmental and social requirements; examine whether current approaches were in line with applicable inter-national law; and recommend an appropriate compliance/enforcement strategy that encompasses both the regulatory and spontaneous approaches.[16]

According to the INCOS report 'Ships, Slaves and Competition' (2000),[17] 'Seafarers in 10 to 15 per cent of the world's ships work in slave conditions with minimal safety, long hours for little or no pay, starvation diets, rape and beatings'. There are companies who send crews over the ocean in ships which are not safe and are devoid of even a minimum stan-dard of equipment for the crew. Those companies usually pay their crews very badly and sometimes not at all. During the last years owners of 'ships of shame' had to adjust their policies to changed circumstances, but they have not yet been driven out of the market. These conditions are flourish-ing in an industry which has much less transparency and public account-ability than other transport sectors. INCOS urges governments of major labour-supplying nations to take on greater responsibility for their seafar-ers. They should review their maritime training and ensure its compliance with the relevant international conventions.

Beyond the strong influence onto working and living conditions played by the flagging out phenomenon, via the breaking of the genuine link between ship, crew and flag state, Gerstenberger (2002)[18] argues that developments in the methods of financing broke the link between shipowner and crew. It is suggested that with the development of the financial market it has become more and more typical that the owners of a ship are no longer identical with its managers. Hence, owners who barely know what is fore and aft in a ship quite understandably tend to see crews quite simply as cost elements or as 'appendages to the technical system ship'. Of course, advance in information technologies and automation makes the onboard equipment a determinant factor for an efficient management of the ship. In the meantime, the overall effect of technical progress has made the role of human labour in terms of monitoring and control even more essential. All of this leads to the deduction that personnel policy has to aim at personnel retention more than personnel rapid turnover and low qualifications.

The role of manning agencies in the recruitment process is significant, emphasising the requirement for adequate regulation. ILO 179 Convention of 1996, on recruitment and placement of seafarers, makes clear that private recruitment services should be regulated. Laws or regulations should be established to ensure that no fees or other charges for recruitment are borne directly or indirectly in whole or in part by the seafarers. The wide variation in the quality of services provided by manning agencies requires close supervision also to avoid means, mechanisms or lists intended to prevent or deter seafarers from gaining employment. Again, higher publicity and transparency is demanded of quality companies and organisations.

In conclusion, what has to be made explicit is that seafaring labour market is not simple or unified neither in terms of regulation nor in terms of living conditions. It rather comprises a diversity of markets cutting across one another and interacting on one or the other in an international environment.

## 4 Safety and culture

The increasing supply of seafarers from non-traditional labour supply areas with insufficient training, infrastructure and doubtful certificates of qualification, the questionable seafarers' commitment with the weakening of the links between shipboard personnel and ownership and language and communication problems among ratings and officers inhibit the safe ship operation by increasing the risk of human error.

Several studies have been conducted on the subject. Roberts (2000)[19] has identified noticeable differences in both the levels and patterns of mortality suffered by British seafarers who were serving in British and foreign-flagged vessels between 1986–95. It would seem evident that British seafarers who take employment in foreign vessels are at greater risk of mortality through work-related accidents. A greater number of seafarers in these foreign fleets also took their own lives, disappeared at sea, or were the victims of homicides.

Concerning the influence of the type of flag and register on the safety issue, Li and Wonham (1999)[4] tested some of the common understandings. First, is that ships registered in open registered countries are more risky than those registered in closed registered countries. Second, is that ships of traditional maritime nations are better managed and safer than those of developing maritime countries. Their study, based on 20 years of records (1977–96) of the 36 principal fleets in the world, confirmed that the worst group in terms of total loss rate is composed by the open registry countries (Liberia and Bahamas being exceptions) and the safety record of developing maritime nations as a group is better than that of TDSCs.

Safety has become an increasingly important issue in shipping also due to ship ageing. The 1970s were economic boom years, but they were followed by relative stagnation. Too many ships were built and demand for new tonnage declined rapidly after the boom ended. As a result, the average age of ships increased steadily. This clearly had implications for safety and environment because old ships tend to be more vulnerable to corrosion and breakdown than new ones, making implementation of high standards more and more essential.

In terms of safety, the IMO has shaped, developed, and put into practice the notion of marine safety and environmental protection not only reflects the organisation's attitude towards these issues but also has affected the formulation of the international conception of marine safety. This is why it is worth mentioning that from the beginning of the 1990s IMO appeared determined to shake off the characterisation of it being simply reactive to marine disasters and it started talking about a philosophy of 'anticipating and preventing' and about 'becoming proactive rather than reactive in improving international legislation'.[20] The efforts to improve ship operations and management and seafarers' standards led to the adoption of a new chapter in SOLAS – Chapter IX – in May 1994 and a revised STCW[21] convention in July 1995. IMO shifted its emphasis from mainly technical aspects to dealing with human factors.

This introduced a more safety-oriented attitude within the industry and was officially seen as instrumental to the concept of safety culture. All the measures in the SOLAS convention come into force in July 2004.[22]

Without denying the positive impact of such initiatives, it is worth stressing that for ship-owners being socially responsible means not only complying with relevant legislation, but also going beyond compliance and investing more than required into human capital and the relations with stakeholders. While various conventions and regulations imposed minimum standards for the operation of shipping companies, the control mechanisms were not always efficient in their mission.

However, quality in shipping is not a matter of the shipowners only.

Neither shipowners nor regulation alone can force or command this price if quality is not demanded by the users of the shipping service by the manufacturers, traders, freight forwarders or final consumers.

With regard to manpower, until now the Maritime Industry has been slow to recognise the impact of leadership and human behaviour on crew retention and reduction of accidents. Nevertheless, it appears imperative that, for ensuring shipping safety, an anthropocentric as opposed to a technocentric approach must be followed in the introduction of all technological innovation in shipping.[23]

The aviation industry first recognised that human error and teamwork failures, as opposed to mechanical malfunction, are major causal factors in industrial accidents more than 20 years ago. It therefore developed human factors training programmes, designed to increase the effectiveness of flight crews, known as crew resource management (CRM).

Recently, CRM has been adopted by a number of other professions including the nuclear and offshore power industries, aviation maintenance, air traffic control, surgical medicine and the Merchant Navy. In the Merchant Navy, the Danish company, Maersk, introduced CRM in 1994. According to Byrdorf (1998),[24] incidents and accidents in the Maersk shipping company have decreased from one major accident per 30 ship years in 1992 (before the introduction of CRM training) to one major accident per 90 ship years in 1996 (after the introduction of CRM training). In addition, thanks to CRM all insurance premiums were lowered by 15 per cent. They attribute this reduction in accidents and incidents to the CRM and simulator training.

Various maritime training centres around the world now offer CRM courses (sometimes under the label *bridge resource management*), and attempts are being made to evaluate the development of key skills, such as shared mental models (Brun *et al.*, 2001).[25]

## 5 Maritime education and training

Maritime Education and Training (MET), manpower skills and culture are interdependent crucial factors to be considered when analysing the evolution of the quality of human resource in the maritime industry. The last ten years have seen an unprecedented amount of both international and European legislation, which is having a profound effect on the standard of seafaring. In particular, the implementation of the revised STCW and the ISM Code imposes major new obligations and responsibilities on national MET systems, as well as on companies and seafarers.

It is possible to identify different clusters of skills (e.g. technical, analytical and social) relevant for human resource in the Maritime Industry. In fact a dramatic change in the Industry that determines new requirements for human resource at international level is the capillary introduction of ICT both on board and ashore. As technology has developed so ships have become technically more complex. As argued above, this has led to the demand for a higher level of technical competence running head long into pressure to further reduce labour costs.

Technical developments would suggest that this trend will continue into the future, albeit to a more limited extent. The whole question of crew size and competence is the starting point for all manning considerations. It is clear from several European Projects carried out under the fourth and the fifth framework programme, that the new requirements under these instruments are being actively addressed within Europe.

Nevertheless, a question mark remains over whether some of the major labour supplying countries will be able to provide seafarers with the skills required to meet the new standards or, if so, whether they will be able to do so in the necessary numbers.[26] Reasons for the inadequate role of the traditional supplier countries in providing manpower for Western Europe can refer to the difficulty in understanding cultural backgrounds and national mentalities as well as overcoming linguistic barriers. The evidence now available indicates clearly that serious problems exist in the ability of multicultural and multilingual crews, along with key shore personnel, to communicate at the level required to enhance and ensure the good reputation of the shipping industry.

Maritime Education and Training Harmonisation research project (METHAR)[27] focused on the development of improved, harmonised and more widely applicable syllabi/curricula for MET students and specialised courses for MET students and lecturers. It also supported the mutual recognition of certificates.

Within the FIT Project,[28] funded by the European Programme ADAPT, 44 out of 122 courses related to maritime; trainees were about 700, equally divided between officers and ratings. The most selected training areas were information technology and foreign languages (English, both general and technical language).

While these skills are generally considered 'basic competencies' for all those trying to enter the labour market, for the professional categories working in shipping (in particular seafarers) they assume the value of very important technical, professional competencies.[29] This is due to the nature of the shipping sector, with its international sphere of action and the spread of information technology that occurred in the 1990s, both of which have affected MET priorities. Likewise, some issues (safety, environment) such as those addressed in relevant international IMO instruments and EU directives, have become the core of syllabuses all over Europe.

In countries where the maritime educational system is still not completely aligned with European requirements, and the entrepreneurial approach towards training is quite reluctant to conceive professional training as a strategic tool to develop human resources, there is less sensitivity to these 'emerging' themes than to 'necessary' competencies (languages and information technology). Another European Project (MARCOM

Project, 1998)[30] was specifically concerned with the problems and practices of Maritime English usage and the training procedures in use. As such its aim has been to contribute to the enhancement of ship safety, environmental protection and stress-free social interaction. More specifically, its main objectives were: to provide an understanding of the significance of communication in the multicultural and linguistically diverse ships of today, and to provide English language instructors of Maritime English with detailed information on the nature of on board use and misuse of language and the types of accidents which can result.

The importance of the English language as a means of effective communication between those on ships and ashore has also been promoted by the Thematic Network on Maritime Education and Training and Mobility of Seafarers (METNET).[31] The general policy objective of METNET is the competitive and sustainable growth in the maritime personnel sector in terms of quality through higher maritime safety and environment protection standards and an enhanced efficiency of shipping operations.

Despite the prominent role that MET could play in shaping shipping working conditions and industry appeal, difficulties and problems for an effective training have still to be solved. In Schröder *et al.* (2001),[32] some reasons for the weakness of MET are highlighted, such as the existence of too different types of MET institutions between countries, a low number of students and lecturers (except in Greece, Italy, Norway, UK), the ageing of lecturers and the small size of MET institutions. Lack of human resources means also absence of research which may help in developing and enriching syllabi and curricula. A striking contrast derives from the generalised lack of resources committed to maritime training and its need of expensive equipment. Concerning training content, important subjects such as economics, logistics and management skills are not taught in sufficient breadth and depth.

Comparing Western Europe MET with MET in traditional suppliers of manpower, some differences arise: in Eastern Europe there are a higher number of applicants than available student places; and in Southern and Eastern Asia there is an enormous demand for student places, due to attractive employment opportunities. Some actions to improve the quality of and harmonise MET are often suggested both in academia and policy-making: exchange of students and staff, networking of MET institutions, MET upgrading.

## 6 Selected works

The choice of the three papers for inclusion in this section derives from their strict complementarity in the issue of human resource in the Maritime Industry. The first contribution 'The Future Shortage of Seafarers: will it become a Reality?'[33] speculates on the number and the origin of manpower, addressing the problem of lack or deficiencies in statistics; the second contribution, 'Finding a balance: companies, seafarers and family life',[34] deals with the issue of the impact of seafaring on family life; the third contribution, 'Seafarers on the World's Largest Fleet',[35] examines the position of labour and its regulation on working on vessels registered in Panama.

In detail, in the first article, Leggate (2004) offers a view of manpower consistency presenting several sources and advocating, at the same time, a stringent need for more reliable and clear figures and statistics on seafarers. She further deals with employment contemporary practice by analysing the regulations and other criteria providing the qualitative framework which underpins seafarers' working lives. The work essentially combines the qualitative dimension with a quantitative analysis. It is borne out of an ILO designed questionnaire distributed to over 70 different government departments, ship-owners associations and unions worldwide during 2001. In addition, it draws on issues derived from a number of case studies on significant maritime nations – The Philippines, India, China, Denmark, and Panama.

In the second article, by Thomas, Sampson and Zhao (2003), it is central to the dramatic impact of the shipping labour model onto seafarers' private lives. The authors draw on in-depth interviews with seafarers' partners in the UK, China and India. The results show how differing lengths for contracts, lack of sufficient leave time, together with difficulties in communication with family make working time harder for seafarers. Some improvements are suggested, such as: shorter trips, continuous employment (rather than employment by voyage) and opportunities for partners and family to sail. Whilst these measures may have financial costs, these can be off-set by improved retention of seafarers and the avoidance of stress-related illnesses. At a time when the commonly held view is of a shortage of seafarers, such steps may be sound company policy.

Finally, McConville (2004) gives a fresh insight into the seafaring labour force through an investigation of the world's largest fleet, Panama. By definition, the fleet is responsible for over 100,000 seafarers. Tighter regulations governing living and working conditions have been established in response to strong international pressure. It is argued that the Panamanian situation is exacerbated by the lack of maritime culture in the country. Panama itself has few seafarers and offers limited maritime services, apart from the Canal. This situation is currently being addressed but is in an embryonic stage. For the seafarers themselves, the regulation is much needed but the major question is one of adequate implementation and enforcement.

The three articles lead to the overriding conclusion of wide diversity in labour market and seafarers' living and working conditions. The industry is vigorously responding to substantial economic and social change, and there is obvious potential for such trends to continue. Such change is reflected strongly in the labour market for seafarers where there must be an increasing awareness of the differing maritime aspirations of many developing countries. It is also reflected in the discriminatory working conditions, making differences between TDMCs and DCs more acute. And third, it has a significant effect on safety for seamen and for the environment as well, due to the destructive consequences of sea accidents.

Against such a dynamic background, this analysis confirms the conviction that regulation at national and international level must be both reshaped and enforced in order to safeguard the industry and the seafarer employed within it. At the same time, within the industry, the strengthening of a shipping cultural identity may help integrate diverging needs for shipowners, manpower and stakeholders, thus making the profession more attractive and the image of the industry more positive and 'friendly'.

## References and notes

1 BIMCO/ISF 2000, Baltic and International Maritime Council/ISF, Manpower update, Seaways 22–4.
2 According to Leggate (2004, see note 3) it is interesting to note however that the Ministry of Communications in China does not consider Chinese seafarers to be a potentially significant source of international seafaring labour. The vast majority of seafarers they train are required to man the substantial Chinese fleet. Having said that, an increasing number of Chinese seafarers are developing a preference for working on foreign-owned vessels, the primary reason being because of the more advantageous wage levels (often 50 per cent higher) which are free of tax.
3 Leggate, H., 2004, The future shortage of seafarers: will it become a reality? *Maritime Policy & Management*, 31, 1, 3–14.
4 Li, K.X. and Wonham, J., 1999, Who is safe and who is at risk: a study of 20-year-record on accident total loss in different flags, *Maritime Policy & Management*, 26, 2, 137–44.
5 Sambracos, E. and Tsiaparikou, J., 2001, Sea-going labour and Greek owned fleet: a major aspect of fleet competitiveness, *Maritime Policy & Management*, 28, 1, 55–69.
6 Lin, C.T., Wang, S. and Chiang, C., 2001, Manpower supply and demand of ocean deck officers in Taiwan, *Maritime Policy & Management*, 28, 1, 91–102.
   Authors recommendations to face the shortage of seafarers in Taiwan are as follows: subsidise education just as the education in teaching profession is; replace the military service with ocean internship of sea-related students; establish training institutes for high-class deck officers and engineering officers; favour alternation between on board and ashore positions.
7 Obando-Rojas, B., Gardner, B. and Naim, M., 1999, A system dynamic analysis of officer manpower in the merchant marine, *Maritime Policy & Management*, 26, 1, 39–60.
8 Moreby, D., 1998, The economics of training ships' personnel in Britain. Paper first delivered in 1968 (Cardiff, UK: Department of Maritime Studies and International Transport, University of Wales at Cardiff).
9 Gardner, B., Naim, M., Obando-Rojas, B. and Pettit, S., 2001, Maintaining the maritime skills base: does the government have a realistic strategy? *Maritime Policy & Management*, 28, 7, 347–60.
10 Leggate and McConville, 2002, Report on an ILO Investigation into Living and

Working Conditions of Seafarers: Case Studies on The Philippines and China, International Labour Office, Geneva.

11 Glen, D., 2003, Regression modelling of the employment duration of UK seafaring officers in 1999, *Maritime Policy & Management*, 30, 2, 141–9.

12 International Labour Organisation, 2001, The impact on seafarers' living and working conditions of changes in the structure of the shipping industry, *Report for the Discussion at the 29th Session of the Joint Maritime Commission*, Geneva (JMC/29/2001/3).

13 Leggate and McConville, 2002, Meeting of experts on working and living conditions of seafarers on board ships in international registers, International Labour Office, Geneva.

14 Leggate, H., 2002, Global seafarer issues, personal communication at *Seminaire Maritime Dest-Inrets*, 21 October.

15 OECD – Maritime Transport Committee, 2001, *The Cost to Users of Substandard Ships*.

16 Morris, H.P., 2002, Globalisation affects today's seafarers, address to apostleship of the sea, XXI World Congress, Rio De Janeiro, 1 October.

17 International Commission on Shipping, 2000, *Ships, Slaves And Competition – Inquiry Into Ship Safety*. The title of the report by the International Commission on Shipping refers to the subject matter of the report. *Ships* refers to international shipping in all its aspects and activities. *Slaves* refers to the thousands of seafarers, mostly from developing nations, for whom life at sea is modern slavery. *Competition* refers to the unequal nature of competition in the international shipping market where non-compliant shipping (substandard shipping) can gain a 15 to 16 per cent cost advantage by not complying with international maritime safety standards.

18 Gerstenberger, H., 2002, Cost elements with a soul, Proceedings of IAME Conference, Panama, 13–15 November.

19 Roberts, S., 2000, Occupational mortality among British merchant seafarers (1986–95), *Maritime Policy & Management,* 27, 3, 253–65.

20 O'Neil, W.A., 1991, World maritime day 1991: a message from the secretary-general, Mr W.A. O'Neil, *Imo News*, No. 3, 7.

21 STCW, Standards of Training, Certification and Watchkeeping, is the main instrument of global quality control. It has been promulgated by the IMO of the United Nations and policed by Port State Control (PSC) in the various member states. The 1995 amendments to the International Maritime Organisation's Convention on STCW have significantly changed the way seafarers are trained and certificated for service on modern seagoing ships.

22 At the heart of the new measures is a new International Ship and Port Facility Security Code (ISPS), which orders governments to employ risk management techniques to ships and port facilities. Each vessel and each port will have to have security plans and a nominated security officer. Ships will need a ship-to-shore security alert system and have to maintain on-board an accurate history of the vessel.

23 Roberts, S, 2000, Op. Cit.

24 Byrdorf, P., 1998, Human factors and crew resource management: an example of successfully applying the experience from Crm programmes in the aviation world to the maritime world, *Proceedings of 23rd Conference of the European Association for Aviation Psychology*, Vienna, October.

25 Brun, W., Eid, J., Johnsen, B., Ekornas, B., Laberg, J.C. and Kobbeltvedt, T., 2001, Shared mental models and task performance: studying the effect of a crew and bridge resource management program, *Rapportserie Fra Prosjekt: Militaerpsykologi Og Ledelse 1*, Norwegian Royal Naval Academy, Bergen.

26 Dirks, J., 1998, *Improving the Employment Opportunities for EU Seafarers: an Investigation to Identify Seafarers Training and Education Priorities*, Federation of Transport Workers' Unions in the European Union (FST) and European Community Shipowners' Association (ECSA).

27 METHAR, 2000, *Harmonisation of European Maritime Education and Training (MET) Schemes in Europe*, Final Report, Brussels: EC.

28 The FIT project was a joint action implemented by the academic, research and entrepreneurial world to boost the Transport and Logistic Industry in southern Italy. The project co-ordinator was IRAT-CNR, Research Institute on Service Industry of the National Research Council, Italy. The aim of the project was to offer a solution to the rising needs of professionalism and flexibility in this sector through innovative, modular training.

29 Carbone, V., 2002, Training needs analysis in transport and logistics: models and tools, A. Morvillo and G. Ferrara (eds) *Training in Transport and Logistics Sector* (London: Ashgate Publishing House).

30 Marcom, 1999, *The Impact of Multicultural and Multilingual Crews on Maritime Communication*, Final Report, Brussels: EC.

31 European Union, April 2000–March 2003, Under the fifth framework programme, has funded, through the thematic programme growth, the thematic network on maritime education and training and mobility of seafarers otherwise called METNET.

32 Schröder, J.-U., Zade, G., Carbajosa, J., Mazières, J. and Pourzanjani, M., 2001, Can maritime education and training (MET) help to increase the attraction of seafaring?, Speech and Paper, Intermediary Project Meeting.

33 Leggate, H. 2004, Op. Cit.

34 Thomas, M., Sampson, H. and Zhao, M., 2003, Finding a balance: companies, seafarers and family life, *Maritime Policy & Management*, 30, 1, 59–76.

35 Based on Leggate, H. and McConville, J., 2002, Report on an ILO investigation into living and working conditions of seafarers: case study on Panama (Geneva: International Labour Office) and Leggate, H. and McConville, J., The economics of the seafaring labour market, C.Th. Grammenos (ed.) *The Handbook of Maritime Economics and Business* (London: LLP), 443–68.

# 7 The future shortage of seafarers

## Will it become a reality?

*Heather Leggate*

## 1 Introduction

There is in the contemporary shipping industry a presumption of an impending shortage of seafarers, specifically officers. This chapter questions as to whether such a shortage will in fact occur through an analysis of the current and potential manning levels and seaborne trade forecasts. It begins with an analysis of the supply position followed by estimates of seafarer demand. It further considers the regional trends and divisions between senior officers, junior officers and ratings, and measures to improve the recruitment in the traditional maritime areas.

The BIMCO/ISF Manpower Update estimates the worldwide supply of seafarers at 1,227,000, comprising 404,000 officers and 823,000 ratings,[1] with the majority originating from a comparatively small number of countries. Over recent decades, changes in economic and commercial activities have been fundamental in the restructuring of the international seafaring labour force. This has been combined with changes in the structure of seafarer employment. At its simplest there has been a relentless decline in the number of seafarers coming from developed countries, due to an appreciable reduction in recruitment and retention. Thus the age structure of this group has become progressively older. The lack of suitable seafarers from developed countries, coupled with a desire to reduce labour unit costs, has created an increasing demand for seafarers from developing countries. These are the main elements which have gradually created a remarkable new concept, that of the seafarer labour-supply country, the majority of which having what can be termed no maritime tradition. Initially the vast majority of these seafarers from these countries were ratings, but they are now supplying a growing number of officers.

## 2 The supply of seafarers

A recent report by the International Commission on Shipping[2] listed the ten top labour-supplying countries (see Table 7.1).

This top ten labour-supplying countries provided 56 per cent of all

*Table 7.1* Top ten labour-supplying countries, 2000[3]

| Country | Officers | Ratings | Total in 2000 | Total in 1995 |
|---|---|---|---|---|
| The Philippines | 50,000 | 180,000 | 230,000 | 244,782 |
| Indonesia | 15,500 | 68,000 | 83,500 | 83,500 |
| Turkey | 14,303 | 48,144 | 62,447 | 80,000 |
| China | 34,197 | 47,820 | 82,017 | 76,482 |
| India | 11,700 | 43,000 | 54,700 | 53,000 |
| Russia | 21,680 | 34,000 | 55,680 | 47,688 |
| Japan | 18,813 | 12,200 | 31,013 | 42,537 |
| Greece | 17,000 | 15,500 | 32,500 | 40,000 |
| Ukraine | 14,000 | 23,000 | 37,000 | 38,000 |
| Italy | 9,500 | 14,000 | 23,500 | 32,300 |
| **Totals** | **206,693** | **485,664** | **692,357** | **738,289** |
| **Grand total** | **403,672** | **823,384** | | |

seafarers in 2000. Of this 56 per cent, some 87 per cent came from the Far East, the Indian subcontinent and Eastern Europe. The primary and obvious point to make is that Filipino seafarers represent some 19 per cent, by far the largest source of supply. China has marginally and perhaps indicatively increased while all the developed countries listed show some contraction. As can be seen, there appears to have been a slight contraction in the total numbers and some minor redistribution between 1995 and 2000.

The International Labour Organization (ILO) commissioned research in 2001 that serves to confirm similar trends but goes further in highlighting the deficit and contradictions within the existing contemporary published information. The ILO research was based on a questionnaire distributed to government departments, shipowners' associations and unions worldwide during 2001, and is included in Table A7.1. Responses were obtained from 73 institutions representing 62 different states. This was further supplemented by a number of case studies of major maritime countries.

The survey and case studies illustrate the lack of definite information on seafarer numbers. Of the 62 flag states questioned, only 38 provided figures for the numbers of seafarers employed on vessels registered under their flag (see Table A7.2). In most cases the figures were estimates. Attempting to quantify the seafarer population is extremely difficult as many countries have no established system to achieve this. Numbers are often based on employment statistics which leaves the problem of determining the number available for employment or *active* population. Furthermore, there are few records of seafarers leaving their profession to pursue other careers.

### 2.1 Labour-supplying countries

As previously stated, it is generally recognized that The Philippines provides the largest number of seafarers. The Philippines' Overseas Employment Administration (POEA) suggests that the number of Filipino seafarers deployed has, in fact, increased by almost 20 per cent over the last five years. The number of registered seafarers in 1999 was 472,225, representing an accumulation of new entrants since registration began in 1988. However, since there is no system for determining when the seafarers cease to be available for employment, this is likely to be an overestimation.

The deployment statistics based on contracts processed by the POEA show a four-fold increase between 1984 and 2000 (Figure 7.1). Given that some seafarers may have more than one contract in a 12-month period (the typical length of contract is ten months) there may be some double counting. However, the POEA estimate that this will be no more than

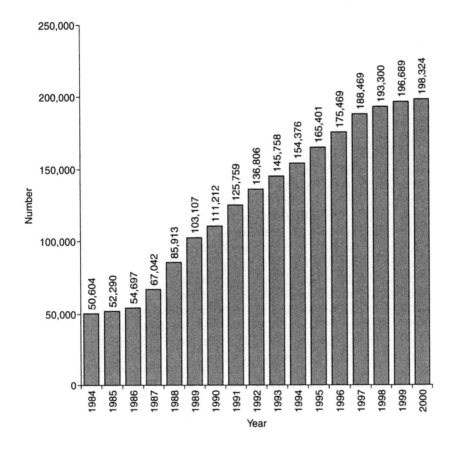

*Figure 7.1* Seafarers' deployed.[5]

5 per cent. The long-run trend from some 50,000 to nearly 200,000 in the level employment over the period serves to highlight the importance of seafaring to the national economy. It is estimated that 60 per cent of the available number of seafarers are deployed[4] which suggests an active population in 2000 of 330,450. This is higher than the BIMCO/ISF estimate.

Despite illustrating a considerable oversupply of ratings, the statistics also highlight considerable potential for officers. The number of new entrants from 1996 has totalled nearly 170,000 and can be analysed by category as shown in Table 7.2.

Table 7.2 shows approximately 33,000 new entrants per year, but with this there is considerable restructuring among the groups. Officers have increased substantially in number by 131 per cent but they still represent less than 1 per cent of the total new entrants in any given year. Ratings, that is deck and engine room, are by far the largest group, but in number and as a proportion of the total have experienced a noticeable decline. The numbers of ratings entering the market have fallen by 8,000 and now represent only 68 per cent of the total. If the last three categories, namely catering, others and luxury vessels, are combined there has been a considerable increase in both their number and proportions, which stood at 16.5 per cent and have doubled to 33.7 per cent during the period. These new entrants must be seen as a strong indicator of future trends, with ratings being under the most intensive competitive pressure. In terms of annual growth, the number of officers is increasing by 23.3 per cent per annum whilst the number of ratings is declining by −7.5 per cent.

An increasing number of international seafarers are being provided by China. The Ministry of Communications estimate that there are currently some 500,000 seafarers in China, with 340,000 (68 per cent) engaged on domestic shipping and 160,000 (32 per cent) on foreign-going ships. Of these, 40,000 are employed on foreign-flagged vessels, which represents a tenfold increase over the last decade. The majority of Chinese seafarers are required to man the substantial Chinese fleet, but for those seafarers who speak good English there is a preference for overseas employment because of better wages. Again, the number of Chinese seafarers is substantially higher than the BIMCO/ISF estimate.

There is currently no accurate published assessment of the numbers of Indian seafarers deployed or active within the industry. The Directorate General of Shipping estimates that there are 24,000 officers, comprising 13,000 deck and 11,000 engine officers. There is a consensus, however, between the government, unions, and shipowners that 70 per cent of officers are employed on foreign flag and 30 per cent on Indian flag vessels. With a preference for foreign flagged vessels due to more favourable tax relief, there is a potential shortage of quality officers to man the Indian flagged vessels. The situation for ratings is rather different. The government estimate the figure at 40,000 although based on union membership the actual total is likely to be higher than this. Given that such ratings are

Table 7.2 New entrants 1996–2000[6]

| | Officers | % | Ratings | % | Catering | % | Other | % | Luxury vessel | % | Total |
|---|---|---|---|---|---|---|---|---|---|---|---|
| 1996 | 118 | 0.34% | 28,484 | 83.06% | 5,072 | 14.79% | 329 | 0.96% | 290 | 0.85% | 34,293 |
| 1997 | 160 | 0.48% | 25,494 | 76.03% | 7,215 | 21.52% | 322 | 0.96% | 342 | 1.02% | 33,533 |
| 1998 | 166 | 0.47% | 25,637 | 72.73% | 8,731 | 24.77% | 457 | 1.30% | 257 | 0.73% | 35,248 |
| 1999 | 226 | 0.68% | 22,670 | 68.23% | 8,801 | 26.49% | 480 | 1.44% | 1,050 | 3.16% | 33,227 |
| 2000 | 273 | 0.89% | 20,864 | 68.29% | 7,213 | 23.61% | 508 | 1.66% | 1,695 | 5.55% | 30,553 |
| Annualized increase | | | | | | | | | | | |
| Total | 943 | | 123,151 | | 37,032 | | 2,096 | | 3,634 | | 166,855 |

attempting to secure one of the estimated 32,000 jobs, there is a substantial surplus of manpower in this area. Ratings have a preference for employment on Indian flagged vessels because of the favourable terms and conditions of employment. However, at present the assumption is that as many as 40 per cent of their number are, of necessity, securing employment on foreign registered vessels in response to the excess supply situation.[7]

In order to obtain a more accurate assessment of numbers, the Directorate General of Shipping is about to introduce a database of seafarers built on training certificates. Any Indian citizen who has obtained a certificate from an Indian college will be given an 'INDOS' number (Indian Database on Seafarers). However, as there is no means of eliminating seafarers who are no longer active from the database, it will merely represent a cumulative total of all those who, at some time in their lives, have received seafarer training in India.

### 2.2 Traditional maritime nations

The trend within the traditional maritime nations has been one of persistent decline in the number of seafarers. In Denmark, the information is considered reasonably accurate based on posts filled since contracts with seafarers are copied to the Maritime Authority. The maritime authorities suggested that the number of active Danish seafarers could be estimated by multiplying this number of posts by a factor of 1.7.[8] Using this formula, the total number of active Danish seafarers was 10,181 in 2000.

There has been, since the establishment of the Danish International Register in 1988, an expansion of the total number of Danish seafarers from 2,578 to 4,173 although their proportion of the total has decreased from 85 per cent in 1988 to 57 per cent in 2000. These positions have been filled by other nationalities, notably Filipinos. Other European seafarers, represented by Scandinavian or EU countries, remain relatively stable throughout the period with 228 (8 per cent) in 1998 and 261 (4 per cent) by the year 2000. Thus during this period of increasing seafarer employment there has been an alteration in the proportionate or percentage change in the employment of other nationalities. This constituted only 8 per cent of the total at the beginning of the period (1988), but by 2000 had moved dramatically taking some 40 per cent of employment opportunities.

Within these changes the proportion of Danish officers has contracted from nearly 97 per cent in 1988 to 80 per cent in 2000. Research revealed that this contraction was in the junior officer positions, which were being filled by non-nationals, particularly Filipinos.

The most severe impact of this redistribution was felt by the Danish ratings. The transition to the new register had a dramatic effect, with a serious and continuing contraction in job opportunities. While the absolute number of Danish ratings increased from 875 to 985 between 1988 and 2000, the proportion actually fell from 74 per cent to 40 per cent.

The Scandinavian and other EU ratings declined in number from 152 to 52 during the period, a reduction from 13 per cent to 2 per cent. The benefit of the new register was, therefore, largely experienced by the other nationalities whose number grew from 151 to 1,388, a proportionate increase from 13 per cent to 57 per cent. This is a reflection of the lower unit cost of the other nationalities compared to the Danish ratings.

In the UK, the number of active officers is known with some precision following a major study in 1997,[9] which has been updated annually by the Centre for International Transport Management at London Metropolitan University. The number of officers has fallen by 16 per cent since 1997 and a further 13 per cent fall is projected over the next five years based on the recent intakes of cadets. Numbers will only remain stable with an intake of 1,050 per annum and the actual recruitment has been 400 to 500.

The research and case studies highlights the fact that the baseline or current position on seafarer numbers is not known with any degree of accuracy. This is indicated by the differences between the ILO study and BIMCO/ISF which, although assimilated from the same sources, in most cases show different results. A total of 35 flag states were common to both studies. The total number of seafarers given for these states was 119 per cent (542,220) higher in the ILO study. These differences are shown in detail in Table A7.2. The ILO research would suggest that the number of seafarers is higher than originally thought, although there is insufficient information to determine the breakdown between officers and ratings. It is not productive to suggest which study is the more accurate, since both rely on data supplied by the flag states themselves. However, what is clear from both studies is that the current position is far from clear.

## 3 Demand for seafarers

Thus far the analysis has focused on the supply of seafarers and the trend towards increasing recruitment from labour-supplying countries and the decline in numbers from the OECD countries. The demand for seafarers is, of course, of major importance for determining a potential shortage. Demand predictions have been provided by BIMCO/ISF based on the expected increases in the world fleet, recruitment and wastage levels, age structures, and manning scales. The demand is estimated at 420,000 officers and 599,000 ratings, which translates into a shortfall of officers of 16,000 and a surplus number of ratings of 224,000. Doubts are, however, expressed about the extent to which many of these ratings are qualified for international service.

In simple terms it is possible to examine the trends in seaborne trade and in the world fleet. The United Nations Conference on Trade and Development (UNCTAD), *Geneva Review of Maritime Transport*, 2002 showed the figures in their latest report.

Table 7.3 shows volatile levels of seaborne trade since 1990 and an

*Table 7.3* Trends in seaborne trade and the world active fleet[10]

| Year | Total (all goods loaded) | |
|------|------------|-------------------|
|      | *Million tons* | *World active fleet* |
| 1990 | 4,008 | 594.7 |
| 1997 | 1,953 | 746.9 |
| 1998 | 5,598 | 764.0 |
| 1999 | 2,668 | 775.3 |
| 2000 | 5,890 | 790.0 |
| 2001* | 5,832 | 804.1 |

Note
* Using provisional data for 2001.

annualized growth rate of 3.5 per cent. In terms of the world active fleet, the annual growth rate over the same period has been 2.8 per cent.

Given the same manning levels[11] the numbers of seafarers would need to grow at 2.8 per cent to sustain the same level of trade. According to the *Manpower Reports*, there has been zero growth in the number of officers and slight decline in the number of ratings over the same ten-year period.[12]

The potential manning problems are exacerbated by the restriction on the use of foreign seafarers by a number of flag states. The ILO study revealed that the majority of states operate some kind of restrictive policy, particularly regarding the master and senior officer positions. Figure 7.2 shows that 8.3 per cent require an all national crew. Authorization for non-nationals is possible in eight countries. There was a requirement in 16 registers for the officers or senior officers to be nationals. It must be

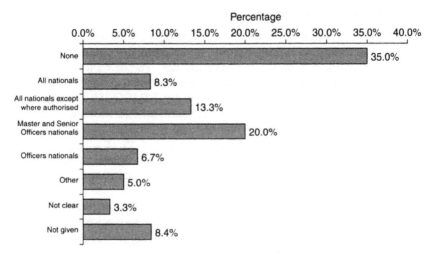

*Figure 7.2* Foreign seafarer restrictions.[13]

stressed that only 35 per cent of countries have no legal restriction on foreign seafarers, and in some of these, non-nationals are rarely found.

The available statistics indicate a potential shortage of seafarers, but will this be the reality? Many associated with the international maritime industry – shipowners, shipmanagers, governments and unions – think it unlikely. Their concern is not for the number but rather for the quality of the future seafarers. Certainly there has been a radical shift in the regional distribution of seafarers, towards the labour-supplying countries. Although the majority of senior officers (46.5 per cent) originate from OECD countries, junior officers are increasingly being provided by the Far East and East and Central Europe. The Philippines is experiencing a growth in the number of officers entering the profession and a relative decline in the number of ratings.

Indeed Figure 7.3 shows that the pattern for junior officers is fast approaching that for the ratings. It would, therefore, follow that such nations will eventually produce senior officers given the falling levels of recruitment in the more traditional maritime areas. If the UK is used as an example, 68 per cent of officers are over 40 years of age.

## 4 Recruitment and retention

The recruitment and retention problems are clearly recognized by the OECD nations, some of which are adopting measures to address the issue. Many European countries have instituted tonnage tax legislation. In the UK this began in 2000 and is linked explicitly to recruitment. For every 15 posts in the effective officer complement for the ships qualifying for the

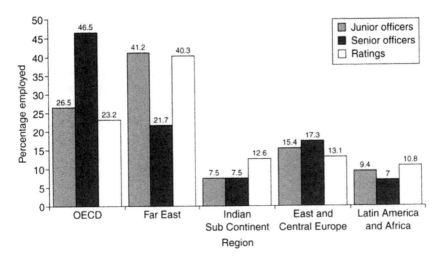

*Figure 7.3* Regional distribution of seafarers, 2000.[14]

tax, the company shall provide the first year of training on a relevant course for not less than one eligible officer trainee.

There is however, an 'opt out' for the company by making a payment in lieu of training calculated by multiplying the number of months when they should have been training by £550. The figures for cadet intakes since the legislation came into force does not yet show any significant increase in the numbers of trainees.[15]

In Denmark, the government has therefore taken positive steps to encourage recruitment into the industry. A substantial marketing effort has been targeted at primary schools in an attempt to heighten awareness of shipping and the broader maritime industry as an interesting and fulfilling career. They have also adopted a dual officer training (engineer and deck) system believing it to be more attractive in terms of transferable skills.

The measures in both countries are relatively new and as yet have had little impact on the recruitment position.

## 5 Conclusions

Discussion with the industry reveal little concern for a shortage of seafarers but a genuine concern for the future number of quality seafarers. Despite the efforts of the traditional maritime nations, there is a definite decline in the number of seafarers from OECD countries which is being counteracted by an increase in those from the labour-supplying countries at junior officer and at ratings levels. Over the coming years we would expect to see a larger number of senior officers from these countries to replace the ageing OECD seafarers. Perhaps their confidence in numbers reflects the fact that the baseline position that we have is uncertain. Indeed, the ILO study suggests that the total supply is in fact higher than suggested by the BIMCO/ISF Manpower Survey. What cannot be disputed is the clear information deficit which exists in this are. Despite the best effort of the studies, the baseline information is inadequate or non-existent. In short, there is a need to devise a simple model for counting seafarers around the world.

# Appendix

*Table A7.1* The respondents

| State | Source |
| --- | --- |
| Algeria | Permanent Mission |
| Australia | Ministry |
| | Union |
| Bahamas | Bahamas Maritime Authority |
| Bahrain | Government |
| Belgium | Union |
| Belgium | Shipowners' Association |
| Brazil | Ministry |
| | Union |
| Bulgaria | Ministry of Transport |
| Canada | Government |
| Chile | Government |
| Columbia | Ministry of Labour and Social Security |
| Croatia | Ministry of Social Welfare |
| | Union |
| Denmark | Shipowners' Association and Ministry Trade and Industry |
| | Union |
| Ecuador | Ministry |
| Egypt | Ministry |
| Estonia | Ministry of Social Affairs |
| Finland | Ministry of Labour |
| France | Government |
| Georgia | Maritime Administration |
| Germany | Shipowners' and Seafarer Associations |
| Greece | Government, shipowners, union |
| Honduras | Government |
| Iceland | Ministry of Social Affairs |
| India | Ministry of Labour |
| Israel | Ministry of Transport |
| Italy | Ministry of Labour and Social Affairs |
| | Confitarma |
| Japan | Government |
| | Union |
| Korea | Union |
| Kuwait | Permanent Mission |
| Lebanon | Ministry of Public Works and Transport |
| Liberia | Bureau of Maritime Affairs |
| Lithuania | Ministry of Labour and Social Security |
| | Union |

*continued*

*Table A7.1* Continued

| State | Source |
| --- | --- |
| Malaysia | Ministry of Human Resources |
| Malta | Government |
| Marshall Islands | Maritime Administration |
| Mauritius | Mission to the UN |
| Mexico | Government |
| Morocco | Government |
| Netherlands | Department of International Social Policy and Information Union |
| Nicaragua | Ministry |
| Norway | Government, shipowners, union |
| Oman | Ministry of Transport and Housing |
| Pakistan | Ministry of Labour<br>Union |
| Panama | Ministry |
| Papua New Guinea | Department of Labour and Employment |
| Peru | Government |
| Philippines | Maritime Industry Authority, POEA, Maritime Training Council |
| Poland | Ministry Labour and Social Policy |
| Portugal | Ministry |
| Romania | Government |
| Russia | Union |
| Singapore | Ministry |
| Spain | Government |
| Sweden | Ministry of Industry, Employment and Communications |
| Ukraine | Ministry of Transport |
| United Arab Emirates | Ministry of Labour and Social Affairs |
| Uruguay | Ministry |
| USA | Government<br>US Coastguard |
| Vanuatu | Maritime Services |
| Venezuela | Ministry |
| Yemen | Union |

*Table A7.2* Numbers of seafarers: comparison ILO and BIMCO studies

| State | ILO study | BIMCO study | |
|---|---|---|---|
| Algeria | 35,605 | 2,635 | |
| Bahrain | 423 | unavailable | |
| Bulgaria | 31,761 | 5,147 | |
| Canada | 900 | 14,633 | |
| Chile | 19,882 | 3,110 | |
| China | 340,000 | 82,017 | |
| Columbia | 12 | 3,455 | |
| Croatia | 6,500 | 19,500 | |
| Denmark | 9,705 | 9,875 | |
| Egypt | 114 | 9,140 | |
| Finland | 10,400 | 10,000 | |
| France | 9,522 | 6,330 | |
| Germany | 11,818 | 14,483 | |
| Greece | 45,363 | 32,500 | |
| Iceland | 200 | 470 | |
| India | 64,000 | 54,700 | |
| Israel | 1,445 | 1,776 | |
| Italy | 12,400 | 23,500 | |
| Japan | 109,644 | 20,913 | |
| Korea | 65,038 | 16,488 | |
| Lithuania | 11,000 | unavailable | |
| Malaysia | 61,830 | 12,671 | |
| Marshall Islands | 17,805 | 40 | |
| Morocco | 3,223 | 2,729 | |
| Netherlands | 14,686 | 11,644 | |
| Norway | 15,216 | 22,200 | |
| Pakistan | 29,655 | 11,808 | |
| Papua New Guinea | 2,530 | 987 | |
| Peru | 3,240 | 1,700 | |
| Poland | 24 | 12,106 | |
| Portugal | 229 | 2,221 | |
| Romania | 1,099 | 10,257 | |
| Spain | 2,243 | 10,000 | |
| Sweden | 430 | 9,600 | |
| United Arab Emirates | 1,080 | unavailable | |
| Uruguay | 2,139 | 1,030 | |
| USA | 15,000 | 15,207 | |
| Vanuatu | 54,145 | 711 | |
| **Total ILO respondents** | **1,010,306** | | 38 states |
| **Total ILO for comparison with BIMCO** | **997,803** | **455,583** | 35 states |
| **Difference** | | **542,220** | |

## References and notes

1 BIMCO/ISF Manpower Update (2000) *The World Demand for and Supply of Seafarers* (Warwick: Institute for Employment Research, University of Warwick).
2 International Commission on Shipping, 2000, *Ships Slaves and Competition*, An enquiry into safety.

3  Ibid.
4  POEA overseas employment statistics.
5  Ibid.
6  Ibid.
7  FUSI estimate.
8  This is based on historical information.
9  United Kingdom Seafarers' Analysis 1997, 1998, 1999, 2000 (London: Centre for International Transport Management, London Guildhall University).
10  World fleet minus surplus tonnage. Compiled by UNCTAD based on data supplied by *Lloyd's Register Fairplay* and *Lloyd's Shipping Economist*.
11  The *BIMCO Report* suggests that there is little scope for further reductions in this area.
12  The 1990 *Manpower Report* shows 403,000 officers and 838,000 ratings.
13  ILO study, 2001.
14  BIMCO/ISF Manpower Update, 2000, Op. Cit.
15  United Kingdom Seafarers' Analysis 2001, (London: Centre for International Transport Management, London Guildhall University) shows 1999/2000 intake at 480, 2000/2001 at 468.

# 8 Finding a balance

## Companies, seafarers and family life

*Michelle Thomas, Helen Sampson and Ming Hua Zhao*

## 1 Introduction

The world's seafarers can be seen as one of the first truly international and global workforces, comprising of individuals from regions as geographically and culturally disparate as Western Europe, Russia, India, South America and The Philippines. Such seafarers work on a range of different vessels, operating different trades, with a diverse range of work conditions. However, one thing that these individuals have in common is that their work necessitates prolonged separation from their home and families, separations that are often characterized by infrequent opportunities for communication. As such, seafaring may be seen as more than an occupation, rather a lifestyle – a lifestyle that involves a constant series of partings and reunions with associated transitions from shore-based life to the unique work environment of the ship. Inevitably, it is a lifestyle that will impact dramatically on both seafarers and their families.

Given the dearth of research on seafarers in general, it is no surprise that little attention has been given to the impact of seafaring on family life or the effect of prolonged absences from home and family on the seafarers themselves. However, the little research that does exist suggests that such separations from home and family may be problematic for seafarers and their families. Research with harbour physicians in Rotterdam identified three main psychological problems among seafarers: loneliness, homesickness and 'burn-out' syndrome. The problems were primarily caused by long periods away from home, the decreased number of seafarers per ship, and by increased automation.[1] Other recent research by the Australian Maritime Safety Association (AMSA) found that seafarers reported the 'home–work interface' to be their greatest source of stress.[2] These problems are not new; a Gallup poll conducted for the Rochdale Committee over 30 years ago reported that problems concerning the separation from partner and family were the most common reasons for seafarers leaving the sea.[3] Such problems may not be without consequence: investigations into suicide at sea have identified marital and family problems as contributory factors to the event.[4]

Whilst seafarers' partners do not have to physically leave their homes and families in the same way that seafarers do, they are, nevertheless, also faced with a relationship characterized by separation and reunion and the constant adjustments these transitions require. Research suggests that such a pattern may affect health, resulting in higher rates of depression and anxiety amongst seafarers' partners than in the general population.[5] As with seafarers, studies of partners highlight the difficulties associated with the transition periods of the work cycle. In 1986, an Australian study of seafarers' wives found 83 per cent reporting some degree of stress when their partners were due home or due to return to sea, with nearly one in ten (8 per cent) reporting taking medication to cope.[6] Nearly half (42 per cent) of the women in this sample felt that their relationship with their partner was strongly at risk due to the seafaring lifestyle and 25 per cent believed that their partner was having, or had had, an affair.

This chapter will focus on the impact of seafaring on seafarers' families. In particular, it will consider the impact of differing conditions of work on seafarers' families and will explore the range of company support available to address and minimize the impact of a seafaring lifestyle. It draws on data collected from two different studies: the Transnational Communities Project (TNC)[7] and the Seafaring and Family Life Study.[8]

## 2 Methods

The data presented in this chapter is drawn from 35 interviews conducted over a 12-month period. Interviews were conducted in the UK, India and China, with a small number of interviews being conducted aboard ship in international waters. Women were identified using a number of strategies, including use of existing Seafarers International Research Centre (SIRC) databases, an advertisement in the National Union of Marine, Aviation and Shipping Transport (NUMAST) Telegraph, contacts made whilst doing shipboard research, and information from shipping companies and trade unions.

Seafaring and family life is a relatively unexplored area and, as such, interview formats were structured in a flexible way, thus ensuring that researchers were not restricted by their own preconceived ideas but would encourage participants to explain things in their own terms, allowing the researcher to explore interesting issues and experiences as they were introduced by the informant. All interviews were conducted in English, with the exception of those with Chinese seafarers' wives, which were conducted in Chinese by Zhao and Chinese collaborators. All interviews were tape-recorded and transcribed verbatim.[9]

The women interviewed for this study were of different ages and points in their lives: some were recently married, some had young children, some had adult children and some had partners who had recently retired. All the women's partners were employed in cargo shipping. The Indian women's partners were both officers and ratings. Chinese and British women were married to officers.[10]

Throughout the text, verbatim quotes are included from the interviews. This gives a vivid account of how respondents think, talk and behalf. Each quote is assigned an identifier to indicate the rank of their partner and the country where the interview took place.[11]

## 3 Cultural context

The women interviewed for this study experienced their lives as seafarers' partners in quite different cultural contexts. The Chinese women participating in the research lived in either Shanghai or Nanyang. Most of the women in Shanghai lived in apartment buildings built and subsidized by their husbands' companies and, hence, in a seafaring community near the port. The living context for the Nanyang wives was different. Far from the sea and the shipping company, these women lived in an environment which was land-oriented and they had little knowledge of shipping or seafaring. The wives of British officers were geographical dispersed and lived in both coastal and inland regions. Many had little previous connection with the sea or shipping. Although occasionally these women were aware of other seafarer's wives who lived in their locality, they usually had little or no contact with each other. In India, the women included in the study lived in very different social, geographic and economic environments. Some officers' wives lived in the highly urban environment of Mumbai. They tended to live in small, sparsely furnished, low-rise apartments on private estates that were generally protected by security guards. They normally relied on domestic help with cleaning and childcare. Their lives were less exposed to public scrutiny than those of their Goan counterparts. In Goa, officers' wives lived in luxurious detached houses with large well cared for gardens and a range of paid helpers including gardeners and maids. They were surrounded by a close community and had much less freedom than seafarers' partners in Mumbai. Nevertheless, many were in paid employment and social attitudes did not seem as constraining as they were in the small village communities that tended to house the wives or ratings. In these small villages, some ratings' partners lived in poverty. They were frequently in debt to their neighbours and families, and where ratings' wives were in paid employment, this was invariably for the little financial reward they could gain. Many wives of seafarers in India had been married by arrangement. Their domestic, as well as their economic and social situations, were, therefore, rather different to those of the British and Chinese partners of seafarers.

# 4 Findings

## 4.1 Working conditions

### 4.1.1 Length of contract

Not surprisingly, length of contract was a significant factor shaping the experience of being married to a seafarer. Contract lengths vary according to nationality and rank, and reflect company employment policies, types of trade and differential labour market values.[12] In China, in order to deal with surplus seafarers, large employers have adopted policies to shorten seafarers' sailing time by as much as half so that 'seafarers take turns to go to sea' (Zhao 2001, personal communication). Such strategies have resulted in Chinese seafarers working for local companies having a sea-time of six months with a corresponding leave period of a further six months. Chinese seafarers working for foreign ship owners can expect to work for one year or longer before they are allowed to take three to four months leave ashore. Indian seafarers experience tours of similar duration, however they correlate more strongly with rank (with officers enjoying shorter contracts than ratings) than with flag or company. As with Indian officers, the partners of women in the UK study worked tours of duty ranging from three weeks to six months, with the majority working three to four months away. For British seafarers, leave period varied from equal time to a ratio of 2:1 (work to leave).

Only those British women whose husbands worked tours of relatively short duration (four weeks or less) reported that they found the length of the period apart acceptable. Regardless of their nationality, the majority of women found such long absences led to considerable problems, including loneliness, while their partner was away and an irreconcilable emotional distance when their partner returned home. As one woman explained:

> It's just an awful long time, you know it's just – they're just away for such a chunk of the year, and every time they come home on a leave for a little bit and then gone, they just seem to be away for an awfully long time
>
> (Third Officer's Wife, UK)

For women whose partners worked longer tours of duty, the difficulties of maintaining an emotional closeness with their partner seemed more apparent. A seafarer's wife in India told us:

> I would enjoy having my husband come home at night, in a way. It's a different life. Not together for months sometimes. Sometimes it gets to you. But I am more used to it than anything, but there's nothing like having the person there. Eight and a half months, and then back

again. As you get old, things change, and you have to get back into the routine. Feelings change, emotions change, ways of thinking change.

(Captain's wife, India)

The difficulties of separation could be exacerbated in the situation where families had young children. Women were aware that their partners were absent for large and significant periods of their children's lives and witnessed the distress of both their partners and their children when husband and child were as strangers to one another. As one seafarer's wife explained:

[It's] too long. And for the children also, they are lost without their father. They want father's love. So it's a problem. That's the main thing when you go for too long ... Yes, he used to feel bad. "Why are they not coming and talking to me?" Sometimes it was like they had not seen their father at all. When they are born, and then he comes home after 7 months or 8 months, they don't recognize who he is. When they were 2 or 3 years, he used to say "Why are they not coming to me?" I said, "It takes time, they must get used to you". Then after they get used to him, they go and play.... Disadvantages in the sense of, the love between the father and the children, it comes less. They don't get to know the father properly, and he's away a very long time at sea. Then he comes home, and stays for a long time with the children. They say, "How long are you home, when do you go away?" Because they didn't know him much.

(Chief Cook's wife, India)

The wives of Indian officers appeared to fare better than the wives of Indian ratings: Indian officers working on Flag of Convenience (FoC) vessels were often in the position to pick and choose contracts so that they might work as little as four months in 12. British seafarers and their partners also enjoyed a better ratio of leave to work than Indian ratings and Chinese seafarers working for foreign companies. Indeed, for the partners of British seafarers, what appeared to be crucial to the experience of their partner being away was not simply the length of the trip, but rather the ratio of work to leave time. A 'one-to-one' ratio of work to leave time was desired by most wives of British seafarers. However, for seafarers' partners from China and ratings' and petty officers' partners from India, their different economic circumstances and seafarers' employment contracts could mean that long periods between tours of duty could, in fact, be filled with anxiety and tension. One seafarer's wife described how:

it becomes too much, with the man at home, they are out drinking, and we are eating. It is too much. When we are at home, we know how to adjust ourselves, but when they are at home as well, it is too expensive. Very high. With the company, there should be a gap of maybe 2 months, or 3 months, at the most. You know how these officers, they have 4.5 months they work and then 4.5 months leave. They know. And still they get wages. But ours, my husband doesn't get wages. He's on completely no wages at all for the 6 or 7 months. That becomes very difficult for us.

(Chief Cook's wife, India)

Most of the women interviewed in Shanghai were redundant workers and solely dependent on the seafarers' wages for financial support. Thus, the reduced wages during the seafarer's leave period could have a significant impact on the financial resources available to the family (at a time when, perversely, the presence of the long-absent seafarers could cause living costs to rise). For Indian women, their partners' contracts were often only for a single tour of duty, so leave periods were often entirely unpaid, resulting in a dependency on savings for the duration of the seafarer's time ashore. Whilst, in some cases seafarers could be recalled by the same company, there appeared to exist a general uncertainty about *when* they would be recalled. Indian ratings without an Indian CDC[13] appeared to find it particularly difficult to get regular work, often resulting in very long periods between tours of duty. For the partners of these seafarers, there existed a conflict between the desire to spend time with their husband and partner and concerns about the economic survival of their family when the seafarer was at home. Such concerns could result in a long-awaited family reunion being fraught with tension and anxiety.

### 4.1.2  Demands on leave time

Leave time may be vital in order for the seafarer to rest and relax after their sea voyage and for couples and families to re-establish bonds and relationships. As noted earlier, for some seafarers, ratios of leave to work time may be very low and, as such, time may be at a premium. The interview data suggests that such shore-based time cannot be considered as uninterrupted free time for seafarers to utilize according to their own wishes and, indeed, figures for leave periods may in fact give a misleading impression of seafarers' time away from the demands and restrictions of the workplace.

Increasing global regulation of the shipping industry, such as the Standards of Training, Certification and Watchkeeping 1995 (STCW '95), has led to increased demands on seafarers to ensure that they meet with industry training standards. Unable to attend courses whilst at sea, seafarers often have no choice but to complete pre-requisite courses during their leave period. Such training courses may be substantial, both in duration and in financial cost to the seafarer and their family. Whilst some courses may last only a few days, this period is increased by, often considerable, travel time, and indeed it was not unheard of for courses to last up to three months.[14] In the context of a leave period of perhaps eight weeks, even relatively short courses could present a significant encroachment on leave time. Both the wives of Chinese and Indian seafarers commented on training demands that infringed on the seafarers' leave period. One Chinese seafarer's partner told us:

> It is an ocean going vessel and the voyage lasts for a minimum of 11 months, normally it lasts for more than a year. Last year he was on the ship for more than 13 months, only came back this April, and a month and a bit later he was called back by the company, to do a 48-day training scheme. The scheme was in Guangzhou. During the 4 months when he was back, he was home for less than half the time. Then he went to the training course, after that he was home for less than a week, then he was called to go back aboard the ship.
>
> (Second Engineer's wife, China)

Ratings' and petty officers' wives in India also reported that their husbands were required to 'report in' to the office upon their return home. The office could be a flight distance away and so further prolonged the separation between seafarers and their families. Senior officers generally did not report back to their offices in person but would wait to be called to return to their ships or would call in when they were ready to return to work.

The wives of British seafarers did not talk about training requirements or demands from the office cutting into leave periods. This may be because more favourable contracts may add any training time to the seafarers' leave period, thus avoiding the loss of time with the family. However, the wives of British seafarers did feel that work invaded leave periods, albeit in a less tangible way than experienced by Chinese and Indian women. The wives of British seafarers reported that their partners often returned home from sea exhausted and stressed and took sometime to unwind and adjust to family life again. Similarly, women reported that their partner's began to make adjustments for their return back to sea as they neared the end of their leave period. Where the leave period was short in duration, couples and families could be left with very little quality time together. As one woman recalled:

> I found it horrendous, he would come home so tired, absolutely zonked out cos [at that time] he was still a second mate and he'd come home absolutely shattered – took him days and days to get over it – and then half way through he would come alive and then be worried about going back to work the fourth week. So you'd always have 2 out of the 4 weeks that were useless.
>
> (Captain's wife, UK)

Thus, regardless of nationality, women in this study found leave periods to be far from unfettered by the demands of work. For Indian and

Chinese seafarers, the need to undergo training to meet with new regulations and international standards often necessitated further periods away from their homes and families. For the wives of British officers, this did not appear to be such a problem, possibly due to better conditions of service which meant that they were compensated for any loss of leave due to training. Other company demands, such as requirements to visit central offices, could also encroach on leave periods. Changes in the industry such as increased automation, decreased crewing levels, increased work load and decreased job security have put pressure on seafarers to put in extra hours to keep their jobs.[15] Such pressures have resulted in increased levels of stress and fatigue[16] and have played a contributory role in a number of maritime casualties.[17–21] These increased hours of work and occupational pressures appeared to expand beyond the confines of the ship to impact on home life. For the wives of British officers, one of the most common problems during the transition period between ship and home was related to the stress associated with the job and the problems their partner had 'switching off' when they returned home from a tour of duty, and the subsequent anxieties in the period prior to rejoining the vessel. Thus, even for those on more favourable contracts, where training time was compensated by additional leave, work concerns could still be manifest and have a detrimental effect on family life.

### 4.1.3 Ship visits and women sailing

The opportunity to sail with a partner was something that was made available to the majority of wives of British and Indian officers. Those women who had sailed were generally very positive about the experience and felt that not only did it give them valuable time with their husbands but that it also led to a greater understanding of their partner's work. Indian women reported feeling that the time spent together aboard improved their abilities to communicate with their husbands and that sailing with their husband allowed them to come to understand that the life of a seafarer was mostly hard work and this prevented them from being resentful or suspicious. As one woman told us:

> Yes. Definitely. I would have felt that understanding may not have been there, the closeness may not have been there in the initial years of marriage, if I had not been able to sail with him. Now that I look back now. That helped me a lot.
>
> (Chief Officer's wife, India)

Where women (British and Indian) were able to take their children aboard to sail, they also felt this had a positive effect on relationships between children and their fathers.

Wives of Indian seafarers employed as ratings were less likely to be able to sail with their husbands.[22] The data suggested that those Indian women who were not allowed to sail regretted that they were not able to do so and seemed to feel shut out of their husband's lives, as the following quote illustrates:

> *Is there anything you think is important that I haven't asked about?*
> Only the seafarers can't take their wives on the ship. That's the worst problem. Otherwise – we are used to it. He must go.
>
> *Why is it so important, do you think?*
> It's like that. For many years my husband has been a shippie. And I should like also to share his job or whatever his life is on the ship. I am a shippie's wife.
>
> (Chief Cook's wife, India)

In China, women have not traditionally been allowed to sail with their husbands. More usually, seafarers' wives have visited their husbands on board when the ship is in port. This was true of the Chinese women interviewed in this study. Women reported travelling long and arduous journeys with their children, and sometimes their families, in order to spend some precious time with their husbands. One woman recalled:

> When the child was small, we have met twice on his ship. Once in 1983 was in Shanghai, he wrote to say when the ship would be arriving and asked me to take the child to Shanghai. Then our child was only 4, we took the train from Huang Shi in Hubei to Shanghai, and it took us 3 days and 2 nights by train. After that, we stayed on the ship for 3 days, then he sent us on the ship bound for home, then his ship left.
>
> (Second Engineer's wife, China)

The opportunity to spend time with each other was valued by the seafarers and their partners alike. However, there was evidence that the drastic reduction of turnaround time had had a direct effect on the length of the wife's visit. In the past, when time in port was longer, the visiting wife could stay aboard with her husband for several days while the ship was operating in port. In contrast, more recently, it appeared to be quite common for a woman to travel for several days from inland provinces, only to be able to see her husband for a few hours (if, indeed, she did not miss the ship altogether). A woman recounted her experience when she visited her husband's ship in 2000:

> He was sailing at sea when our child was born. Our child is 8 months old now. Three months ago, when his ship was calling Qingdao, I took the child and my parents to visit him there. He missed us very much and said that he would be happy if he could only have a look at the

baby. Otherwise, he would have to wait for another few months and when our child could become 1 year old. So I agreed. I had a painful arm, I had to take my parents to help me on the way. It was a long journey, from Nanyang to Qingdao, 29 hours by train. By the time we arrived there, it was already 5 o'clock in the afternoon and the ship would be sailing at 10. We met for only 4 hours. I was really sad ... I wish he could have stayed longer. But the ship had been loaded and unloaded so quickly that I had to take my baby and leave the ship after only meeting with him for 4 hours!

(Second Officer's wife, China)

The accounts of British and Indian women who had had the opportunities to sail with their husbands reflect the beneficial effects of this opportunity for both seafarers and their partners. Indeed, the efforts made by Chinese women to visit their husbands, even for very short periods, illustrates the importance of physical contact, however brief, for couples and their families. For women, spending time with their husbands aboard served to bridge the gap between ship and shore life and to facilitate support and understanding between couples. Women could have an insight into the occupational world of their husband whilst simultaneously reducing or avoiding the lengthy separations which could be so detrimental to relationships.

### 4.2 Company support

The level of company support varied considerably amongst the women in the study. Chinese women appeared to have considerably more company support than their Indian and British counterparts. The most clear illustration of this can be seen in the example of the 'Seafarers' Wives Committee', which was introduced by the Party Committee through the trade union of the shipping company. The chief objective of the Committee was to 'unite seafarers' wives at home front and to provide support for seafarers at sea' (Zhao 2001, personal communication). As a woman in charge of the Seafarers' Wives Committee explained:

The Committee commits itself to help the seafarers' families. We have a tradition here. Any seafarer returns from the sea for leave, one of the first things he does is to report to the Committee. He would drop in and say hello to us. And, he would say, 'Hi, I am home for 2 months. Please don't hesitate to let me know if you need any help'. In this case, we would know who is at home and who isn't. Then, when seafarers' wives need help, for instance, when they need to move house, to buy coal or do other physically demanding tasks, we would give some of the men a ring and ask them for help. Oh, yeah, they are always happy to help, because they know their wives may need help when they are away sailing at sea.

Aided by the close physical proximity of seafaring families to each other, the Committee played a crucial role in organizing the seafarers and their families together and formalizing a mutually beneficial relationship which otherwise could only be realized through the informal means of seafarers' social networking. The Chinese wives recognized the work and contribution of the Committee in promoting the welfare for seafarers and their families. As one woman recalled:

> His work unit has been good to me, especially when I was sick. In 1988, I somehow adopted hepatitis and our son was only 7 years old. I phoned his company, asking if they could allow him to leave this ship and come back to look after me. They told me that it might take him a while to return, but they offered to send a person to help me. Although I declined their offer, because hepatitis is a very infectious disease and I didn't want anyone to pick it up, I have been touched by their kindness. Then, in 1999 when I had a major surgery, the company phoned me several times to send their best wishes. They also gave me some money and bought me fruits when I returned home from the hospital.
>
> (Chief Officer's wife, China)

However, at the same time, these women also expressed disappointment with the company as, as a result of the country's economic reform, such an important service has been reduced in recent years. As one woman told us:

> The company calls us as Haifuren (seafarers' wives) and we also refer ourselves so. I believe that haifuren's role is very important, but the work unit doesn't pay us much attention now. They used to organize parties or other gathering occasions for us, but they no longer organize such activities now.
>
> (Captain's wife, China)

The importance of the clustering of seafaring families to the success of the Committee can be seen when the situation of seafaring wives in inland areas is examined. The geographical distance between families prevented the 'Wives Station' to function effectively or in the same way as the Committee in the port cities where most seafarers and their families live in the same apartment buildings. A Chinese seafarer's wife explained:

> Well, I met them only once, it was when they came to organize the Wives Station. Other than that, my contact with them is mainly by phone or post. They sometimes send forms for us to fill in, such as the medical form for him, the family planning form, their annual thanks letter, etc. They were always very kind whenever I phoned them ... but I don't think that I can rely on them for practical help. I have to depend on my own family, my parents for help whenever I have any problem.
>
> (Second Officer's wife, China).

In this case, seafarers' wives were more dependent on their informal social networks, especially their own parents, for practical support.

In contrast to the experience of the Chinese seafarers' wives, company contact was reported to be low by several of the wives of British officers. Approximately one third of these women said they had little or no contact with the company that employed their husband. About a third of the UK seafarers' wives had had reason to contact the company for the husband to be brought home in a family emergency, ranging from a parent's stroke to the illness of daughter. However, many women had had negative experiences in relation to their partner's company. In extreme cases, two seafarers had been made unexpectedly redundant and informed by letter after many years of service for the same company, another woman found, from reading a national newspaper, that the ship her husband was on was caught in cross-fire during the Gulf War and a third seafarer was never paid over £20,000 salary owed to him. Less extreme was the frequently expressed complaint relating to the unpredictability of work schedules. Difficulties finding reliefs was reported as extremely stressful for seafarers and their partners and families and led to an inability to plan, from small events like trips to the theatre to family holidays. The uncertainty associated with transitions from work to home and from home to work could be a particular source of tension for seafarers and their families. As one woman explained:

> Like I was, that's another thing because it disrupts my lifestyle because I work full time when he's away and he'll come home unexpected now and I wasn't due for holidays until September cos I was hoping to book a couple of weeks off when he comes home in September. So now he won't be home in September and of course I'm saying 'is there any way you can give me a week off cos he's only home for a week?', you know.
>
> (Captain's wife, UK)

The experiences of the wives of Indian seafarers varied according to the rank of their husband: senior officers' wives appeared to have good company support and access, rather similar to that reported by the wives of British officers. However, for the wives of Indian ratings, the situation was quite different. Some women reported that they felt they were deliberately kept in the dark by their husbands' companies and sometimes they were just not kept informed at all. As one wife recalled:

> *If there was an emergency at home, do you know how to contact him?*
> Yes. I try to contact the office. Last time, when I lost my mother, they didn't know where he was. We sent a fax message to them, but they didn't contact [him]. [. . .]
>
> *How long was it, from when you lost your mother to when he came?*
> It was 28 days. My brother came from abroad. But my husband didn't get the message.

*So you knew how to contact him, but it didn't work. And you don't have any direct contact with the ship?*
No.

(Motorman's wife, India)

The experience of the Chinese seafarers' wives in terms of company support largely reflects the specific historical and cultural context.[23] However, regardless of the origin – it was clear that this support was valued by the women and appeared to lessen the impact of having a partner who was away from home for long periods. More support and contact from the company was often mentioned as a means of improving the welfare of seafarers and their families by both Indian and British seafarers' wives. In particular, it was felt important that partners should feel that they could get in touch with their husband in the event of an emergency and that the company should keep families informed of their partner's ship's movements. Several British seafarers' wives reported that they would welcome any efforts by their husband's company to put them in touch with other seafarers' wives and families. Efforts to reduce the uncertainty regarding dates as to when their partners were due to return home or join ships were also steps that women felt would be welcomed.

### 4.3 Communication

The separation from family and home has been found to be one of the most significant factors contributing to stress amongst those in offshore industries.[24] Contact with home can be particularly important at times of ill health of family members when stress levels at sea can rise dramatically.[2] Advances in communication technology have considerable significance in the lives of work-separated couples[25] and in maintaining relationships with the family and shore-based life.[26] Indeed, reduced frequency of contact can lead to relationship decline and eventual breakdown.[27]

It is not surprising that for all women, regardless of nationality, communication with their husbands and between ship and shore was of considerable significance. Communication was important for a number of reasons; to allay fears, to maintain close relationships, to improve seafarers' morale, to relieve stress (on board and at home) and to maintain relationships with children. As one woman explained:

It's OK for me, we don't have kids. But others, it is very difficult for a father to leave the child and go. When they go they really feel bad. They don't like to leave the family. They wait for the letters from the family, I have seen them. Specially when they go to port, they come running, "My letter has come". And they are happy, when they receive their letters. The person who does not receive his letter – my husband always tells me, "Write something and send it to me". When they are over there, they keep on thinking about the family. So when you read something again and again, it makes you feel a little better.

(Chief Officer's wife, India)

For Indian women who had had arranged marriages, ship-shore communication could be vital, not just to sustain the relationship, but for the couple to actually get to know each other. One woman told us:

I think – because you know he keeps on ringing me up from every port, every port there is not any money matters for him. He will just keep talking. I say "It is becoming expensive", but he will say no, no. "You keep on talking to me". Then he will call his mummy, call his father. He kept on talking to them. After that, I came to know. Because initially, since we didn't know each other, you always have that feeling, because you don't know the person as such.

(Chief Officer's wife, India)

Seafarers and their partners reported utilizing a wide range of forms of communication, from conventional letters to satellite and mobile phone calls and e-mail. Advances in communication technology were heralded as quite life-changing to this group. Increasing access to e-mail and to cheaper international phone calls via cell net phones served to expand opportunities for communication considerably. As one woman explained:

Before, we contacted each other by letters. Letters were our spiritual food then. We wrote to each other a letter every month. Occasionally we made phone calls. But phone calls were too expensive at that time, so I would have to prepare well what to say before dialling the number. Now we use the phone and mobile to keep in touch.

(Captain's wife, China)

However, whilst advances in telecommunication technology were undoubtedly advantageous to seafarers and their partners, access varied considerably. All wives appeared to use a combination of letters and phone calls, with British, and occasionally Indian, women reporting using e-mail to stay in touch with their partners. All couples were, in fact, very reliant on modern communication technology for contact with their partners. All of the British couples (where the seafarer was currently serving) used cell phones to communicate (both nationally and internationally). This is considerably higher than the national rate of mobile phone ownership.[28] In addition, nearly half of the British couples reported having e-mail facilities at home. A much higher proportion than those wives of Indian and Chinese seafarers.

Those women who had access to e-mail were very positive about its effect on their lives and their relationships. One woman told us:

It's [e-mail] absolutely wonderful because whereas before I'd say "Oh bloody hell the girls – they've pissed me off!" or something like that.... Now he can say "well what have they done now?" Whereas before I'd have had to bottle it all up and you might put it down on paper but when you do that it isn't anything like the day that you've gone through. Maybe by the time he's come back you've got it all resolved but its better to be able to share it there and then.

(Captain's wife, UK)

E-mail and telephone conversations allowed wives to keep their partner informed of small day-to-day events that might not be reported in a letter or mentioned on their return home. The frequency and style of e-mail and telephone conversations was reported to be vital in managing the transition from home-to-work and work-to-home and in linking the two domains so that movement between the two was less problematic. Those British seafarers working coastal routes could often call home using shore-based mobile networks, at dramatically lower costs than satellite phone calls. Weekly telephone communication was not uncommon for those British seafarers working in these conditions. Emotional needs were met by frequency of telephone calls, but such contact was also valued for its practical implications, allowing seafarers to take part in and respond to household and family decisions such as queries over house insurance and decisions relating to children's wellbeing. Two women told us:

Then there were no telephones, we wrote to each other. At the shortest there were a letter every 1.5 months, and at the longest it was 2–3 months. Now we have telephone in the house, it is a lot convenient. He says that he bought a telephone card there and that it is cheaper if he calls us from the port. He cares about his family, although he can't help me much, only letters and telephones, it gives you a psychological comfort.

(Second Engineer's wife, China)

*How do you feel now because he is working on coastal routes?*
I feel like he is working somewhere in [local town] because we keep in touch all the time by phone.

(Chief Engineer's wife, UK)

However, technological advance was seen as a double-edged sword by some of the women. Chinese women reported still finding it quite expensive to phone internationally and opportunities for seafarers to call home were found to be impeded by the drastic reduction in turnaround times and the development of ports in areas remote from town facilities, leaving seafarers without the time or the facilities to call their families.

The accounts of women interviewed highlighted the crucial importance of communication to seafarers and their families. Communication allowed relationships to be developed and sustained, often over lengthy absences, and provided opportunities for couples to provide mutual support and for the seafarer to feel an active part of the family and household through participation in every-day events and decision making. Despite technological advances, the data suggested continued discrepancies in access to telecommunication facilities. Some wives of Indian ratings reported only recently having a domestic telephone installed at home, whereas it was not uncommon for wives of Indian officers to have e-mail facilities at home and all of the wives of serving British officers reported having a household and cell net (mobile) phone from which contact could be made in both national and international coastal waters. Many couples in these studies were fortunate in that seafarers held senior ranks and had access to shipboard telecommunication facilities (such as e-mail) and salaries that allowed the financial costs associated with communication to be less than prohibitive. However, this may not be the case for seafarers of different ranks and nationalities[12] and indeed, did not appear to be the case for the wives of Indian ratings interviewed for this study. Access to cheaper (or free) communication was frequently mentioned as a means of improving the welfare of seafaring families and reducing the negative effects of a seafaring lifestyle on family life.

## 4.4 Hidden costs

Seafarers are often seen as high earners and in India and China being 'dollar-earners' gave seafarers and their families a 'wealthy' status within their communities. In the UK, the tax-benefits of working in international waters can often mean seafarers have a higher disposable income than many of their peers. However, seafaring is not without its costs. The emotional cost to seafarers and their families may be immeasurable. However, in addition, this research suggests that whilst, for many, the relatively high salary is an incentive to work at sea, the very nature of the seafaring occupation can place additional financial burdens on seafarers' families which are not faced by those with shore-based jobs.

Perhaps the most significant financial cost for seafaring families, regardless of nationality, was the cost of communication. As outlined previously, in the absence of any opportunity to be physically in each other's presence, communication between ship and shore could be vital to seafarers and their families. Indeed, for many seafaring families communication was of such importance that the financial implications of 'staying in touch' were often disregarded. As one wife told us:

> Sometimes he calls me twice a day. Sometimes – well, it depends. If he gets a port, the first thing he will do is call me. Sometimes he does satellite calls also. He really doesn't think about the money. He talks to me and I talk to him, and we feel good about each other.
>
> (Chief Cook's wife, India)

In addition to the cost of the communication, the cost of purchasing equipment to communicate could also involve large initial financial outlay. Of the British couples, approximately half of those households where the seafarer was currently working had access to e-mail at home.[29] It was not clear whether access was solely the result of the need to communicate with an absent partner or whether the equipment and Internet connection would have been purchased regardless. However, the initial cost of purchasing computer hard and software and the ongoing cost of Internet connection can be considerable. Other couples in the study reported buying fax-machines to be installed in their homes as a cheaper alternative to satellite phone calls. It is unlikely that such equipment would have been purchased if their husbands had been in shore-based employment.

Other costs associated with seafaring reflected the (sometimes vast) differentials between employment contracts and conditions of service. Some wives of Indian seafarers reported that their husbands occasionally had to pay, quite considerable, sums of money in order to secure a contract for a single voyage. Where seafarers were employed on single voyage contracts then money had to be saved to cope with periods of unemployment between contracts, and, in the case of Indian seafarers, costs of training and travelling to training venues often had to be met by the seafarers themselves. Whilst the Chinese seafarers in this study were paid during their leave period, this was at a rate considerably reduced from their salary during sea-time. Thus, seafarers and wives had to manage money to cope with these fluctuating salary levels and periods where there was no income at all. As one woman explained:

> *So, you're really careful when he's at sea.*
> Yes. Very careful. We have to save. If something happens, something comes up, and you are not able to cope, sickness or something, or accident takes place, how are you going to manage? So, I keep on saving a little bit. Whatever expenditure I have, and then what I have over I keep aside.
>
> (Chief Cook's wife, India)

Even for those seafarers on more favourable contracts, if their wives wished to visit a ship or join a ship, the cost of travel had to be met by the family.

The nature of a seagoing career also impacted on women's own employment choices. Many of the British women in the study reported that their own participation in paid employment was a crucial factor in enabling them to cope with their partner's intermittent absences. This also appeared to be true of the wives of Indian officers who were all working for reasons other than economic gain. However, paid employment could also be problematic for the wives of seafarers. British women talked of the difficulties in arranging leave to coincide with their partner's leave periods and, in one case, one woman chose not to work for this reason. Other women took more flexible (and hence often lower paid) jobs that allowed them to work reduced hours when their partner was home. Both Indian and British seafarers' wives talked about giving up their own paid work in order to sail with their husbands. As one wife told us:

> I didn't work, because after I got married it was always going to the ship, coming for a short holiday, going back again to the ship, then a holiday. So I never had the opportunity, or perhaps never myself to try to work. That work would confine me to stay separate from my husband. So my option was to sail, so I never worked.
>
> (Captain's wife, India)

Other British women chose not to work as they felt this would have a detrimental effect on their children who already had to cope with an intermittently absent father. These women felt with a father at sea, their children needed the consistency and security of their mother at home. Thus, seafaring may also impact financially on the household in the loss of potential earnings of the female partner.

## 5 Conclusion

It may be argued that seafarers' family and home life are peripheral to the workplace and, therefore, not of company concern. However, a consideration of, and interest in, the family dimension is, in fact, sound company policy. The problems of the retention and recruitment of well-trained seafarers is a matter of global concern.[30] Data shows the stress associated with separation from family is significant for seafarers[2] and that separation from family is one of the most important 'stress' factors influencing a decision to reduce planned sea service.[31] Indeed, in their accounts for the Seafaring and Family Life study, many seafarers spoke of colleagues leaving the sea due to pressure from their partners and families, and the difficulties they themselves experienced being separated from home and loved ones. High staff turnover has significant financial implications for companies, and indeed in the current and projected future labour market, companies may face considerable difficulties replacing exiting seafarers with crew of a necessary high calibre. These issues aside, company retention of existing seafarers is vital to stable, effective and safe crewing. Maintenance of regular crew has important implications for safety, teamwork and effective communication within the ship setting. Increased duration of employment within a company fosters company loyalty and allows an awareness of specific company policies to be developed and crew to be effectively trained according to company requirements. Thus, maintaining a stable crew directly effects and promotes improved crewing and safety standards.

Anxieties about family and loneliness caused by prolonged separations and lack of opportunity for contact can also impact on seafarers' work performance and this may have significant repercussions on safety within the work environment.[32] Indeed, even where there are no perceived problems in family relations, the emotional deprivation associated with prolonged absences from partner and loved ones can lead to psychological deterioration and increased rates of emotional tension which in turn may lead to increases in stress, emotional alertness and aggression, threatening individual and workplace health and safety.[33] In the context of the high number of accidents attributable to 'human error'[8,34] such factors should not be disregarded.

Intermittent separation from family and home may be seen as an inherent and unavoidable feature of seafaring. These interviews suggested that absences and separation do have a considerable impact on seafaring families. However, the data also showed that this impact was neither uniform nor indiminishable. Rather, the conditions of service and degree of support from the company can considerably effect the experience of seafarers and their partners.

Experiences of seafarers' partners in these studies varied. However, this was not just by country. The, sometimes dramatically, different experiences of the wives of Indian seafarers could be seen as directly related to the rank of their husband. For those Indian women married to senior officers, the impact of the seafaring lifestyle was lessened due to their partner's more favourable conditions of service: shorter trips, better company access and support, opportunities for partners to sail, ready access to rejoin vessels and higher salaries, allowing better access to communication technology. Indeed, the experiences of the wives of Indian officers was, in fact, more similar to that of the wives of British officers, than wives of Indian ratings. The lives of Chinese seafarers' wives were also varied, reflecting whether their partner was employed by a national flag or a FoC and their location, whether in a traditional port or coastal city, such as Shanghai where they lived in close proximity to other seafaring families and received company support in the form of the successful 'Seafarers' Wives Committee', or in an inland region where they were geographically separated from other seafaring families and, as a consequence, received little effective company support.

However, despite these differences and regardless of nationality or partner's rank, the effect of varying conditions was the same. Shorter trips were found to be advantageous for all involved, allowing family and couple relationships to be developed and sustained. For the wives of Indian ratings and Chinese seafarers working on foreign flagged vessels, leave periods were often strikingly short in comparison to the many months (ten or more) spent at sea. Leave periods were often encroached on by training requirements introduced by global regulations. For many women in this study, periods between their partner's tours of duty were tainted if not spoiled altogether by the anxieties associated with the sudden reduction or indeed cessation of the seafarer's salary into the household. Tensions could be increased where contracts were for single voyages only and there was no assurance of future employment. Such unpredictability made it difficult for seafarers and their families to budget and manage household finances and could lead to tension within relationships. The wives of British officers reported that increased pressure in the workplace meant that their partners were often stressed for a considerable portion of their leave period, again affecting quality of time at home.

Wives were very positive about the opportunity to sail with their husbands and felt this led to an increased understanding of their partner's work environment;[35] however, this was an opportunity that was often restricted to the wives of officers. In the absence of physical contact, communication took on an increased significance, however access was highly variable and communication was not without a financial cost. The financial implications of purchasing communication equipment (household telephones, cell net phones, fax-machines and personal computers for Internet and e-mail access) and the on-going costs of making contact were considerable and sometimes prohibitive.

Interviews with seafarers' partners suggest that there are a number of steps that can be taken to reduce the impact of seafaring on family life. These can be very effectively undertaken by companies, with any financial costs off-set by better retention of expensively trained staff who might otherwise leave the sea or be subject to stress-related illnesses. In particular, this study showed that efforts should be made to ensure:

- shorter trips (preferably no longer than four months);
- paid leave of a comparable duration to sea-time;
- continuous employment rather than employment by voyage;
- training time to be added on to leave period;
- opportunities for partners (and where possible, children) to sail;
- improved access to cheaper communication;
- increased contact between seafarers' partners and seafarers' employers; and
- opportunities for seafarers' families to make contact with each other.

These changes will benefit seafarers' partners and families, and also seafarers themselves, with direct positive consequences for their employers. Partners and families are a neglected but vital part of the success and the sustaining of the shipping industry. As one seafarer interviewed for the Seafaring and Family Life study noted:

> And it's like ... a seaman's life is all about freedom isn't it? He comes, he goes, he travels, but you've still got to have a base. Without a base you've got nothing.
>
> (Captain, UK)

If companies wish to employ stable, content and above all, safe crews, then they could do worse than to give some attention to seafarers' families.

# References and notes

1 Agterberg, G. and Passchier, J., 1998, Stress among seamen, *Psychological Reports*, 83, 708–10.
2 Parker, A.W., Hubinger, L.M., Green, S., Sargaent, L. and Boyd, R., 1997, *A Survey of the Health, Stress and Fatigue of Australian Seafarers* (Canberra, Australia: Australian Maritime Safety Authority).
3 British Parliamentary Papers, 1970, *Committee of Inquiry into Shipping [Rochdale Report] CMND 4337* (London: HMSO).
4 Roberts, S., 1998, *Occupational Mortality among British Merchant Seafarers: A comparison between British and Foreign Fleets 1986–1995* (Cardiff, UK: Seafarers International Research Centre, Cardiff University).
5 Parker, A.W., Clavarino, A. and Hubinger, L.M., 1998, *The Impact of Great Barrier Reef Pilotage Work on Wives and Families* (Canberra, Australia: Australian Maritime Safety Authority).
6 Foster, D. and Cacioppe, R., 1986, When his ship comes home: the stress of the Australian seafarer's partner, *Australia and New Zealand Journal of Family Therapy*, 7, 75–82.
7 Kavechi, E., Lane, A.D. and Sampson, H., 2002, *Transnational Seafarer Communities* (Cardiff, UK: Seafarers International Research Centre, Cardiff University).
8 Thomas, M., 2002, *Seafaring and Family Life* (Cardiff, UK: Seafarers International Research Centre, Cardiff University).
9 Chinese interviews were transcribed in Chinese and translated to English.
10 In the British data there was an over-representation of women married to Captains and Chief Engineers, reflecting the increased likelihood of those in these groups to respond to the recruitment advertisement.
11 Throughout this chapter, seafarers will be attributed the male gender and their partners, female. However, the authors recognise that seafarers may be female and have male partners, and that both male and female seafarers may choose same-sex partners. It is also recognised that partners, may, in some cases not be married and hence not be 'wives' or 'husbands'. However, for the purposes of this chapter, the choice of language reflects the characteristics of those participating in the studies from which this data is drawn.
12 SIRC, 1999, The Impact on Seafarers' Living and Working Conditions from Changes in the Structure of the Shipping Industry. Report prepared for the International Labour Organisation by the Seafarers International Research Centre, Cardiff University, UK.
13 Indian Seaman's Book.
14 During fieldwork in Shanghai, Zhao spoke with a group of seafarers taking a three-month course in language and STCW training at a marine college. Seafarers who stayed on campus were not allowed to visit their families until the weekend and for some, who lived in other cities, this meant separation from their families for the entire duration of the course.
15 Collins, A., Matthews, V. and McNamara, R., 2000, *Fatigue, Health and Injury among Seafarers and Workers on Offshore Installations: a Review*, SIRC Technical Report Series (Cardiff, UK: Seafarers International Research Centre, Cardiff University).
16 NUMAST, 1995, *All in Good Time* (London: NUMAST).
17 MAIB, 2002, Collision between a tank barge and vessel alongside, *MAIB Safety Digest 1/2002, Case 7*, http://maib.dft.gov.uk/sd/0102/11.htm.
18 MAIB, 2002, Grounding of general cargo ship with pilot on board, *MAIB Safety Digest 1/2002, Case 11*, http://maib.dft.gov.uk/sd/0102/15.htm.

19 MAIB, 2001, Collision with the Nab Tower, *MAIB Safety Digest 1/2001, Case 7*, http://maib.dft.gov.uk/sd/0101/11.htm.

20 MAIB, 2001, Grounding of a container feeder, *MAIB Safety Digest 2/2001, Case 2*, http://maib.dft.gov.uk/sd/digest/06.htm.

21 MAIB, 2001, Bad dream becomes a reality – middle watch grounding, *MAIB Safety Digest 3/2001, Case 1*, http://maib.dft.gov.uk/sd/0301/05.htm.

22 One company that employed several of the Indian seafarers whose wives were included in the study allowed the wives of any ranks to sail, subject to available (suitable) living accommodation and safety regulations.

23 In China, employees refer to their employers, such as schools, factories or hospitals, as *danwei* or the work units. Under the planned economy, the work unit was responsible for providing its employees with both their wages and virtually all the social services such as childcare, children's education, medicine, pension and even incurred cost for funerals. Whilst economic reform has drastically restructured the institution of the work unit in the last 20 years, the work unit still carries far more weight in shaping individuals' work and life than most employers in the Western economy.

24 Sutherland, K.M. and Flin, R.H., 1989, Stress at sea: a review of the working conditions in offshore oil and fishing industries, *Work and Stress*, 3, 269–85.

25 Robertson, J., 2001, The Effects of Short-term Work-related Separation on Couple Relationships: Army Servicemen, Merchant Seamen and Commuters Compared, MSc Thesis, Institute of Psychiatry, Kings College, London, UK.

26 Davies, A.J. and Parfett, M.C., 1998, *Seafarers and the Internet: Email and Seafarers' Welfare* (Cardiff, UK: Seafarers International Research Centre, Cardiff University).

27 Argyle, M., 1990, Social relationships, *Introduction to Social Psychology*, edited by M. Hewestone, W. Stoebe, J. Codol and G. Stephenson (Oxford, UK: Blackwells), 222–45.

28 From 1990–2000, 44 per cent of households in Britain had at least one mobile phone. National Statistics, 2001 *Social Trends*, 2001, edn no. 31 (London: The Stationery Office).

29 This is considerably higher than statistics for the general population, which showed that, in 1999–2000, 38 per cent of households in the UK had a personal computer. National Statistics, 2001, *Social Trends* 2001, edn no. 31 (London: The Stationery Office). Data was not available on the percentage with Internet access. However, it is likely that when this is taken into consideration the figure would drop considerably.

30 BIMCO/ISF, 2000, *BIMCO/ISF 2000 Manpower Update – The World-wide Demand for and Supply of Seafarers*, http://www.marisec.org/2000Update/2000update.htm.

31 Telegraph, 1999, Research bid to cut officer 'wastage', *NUMAST Telegraph*, 3 November.

32 Research with airline pilots has suggested that domestic stress and other major life events may have a detrimental effect on a pilot's judgement and wellbeing. McCarron, P.M. and Haaksoson, N.H., 1982, Recent life change measurement in Canadian forces pilots, *Aviation, Space and Environmental Medicine*, 53, 6–12. The importance of the spouse as a social support system and in enabling the pilot to cope with stress has been acknowledged by the Aviation industry, along with the specific problems associated with a marriage where one partner is frequently absent. Karlins, M., Koh, F. and McCully, L., 1989, The spousal factor in pilot stress, *Aviation, Space, and Environmental Medicine*, 60, 1112–15.

33 Horbulewicz, J., 1978, The parameters of the psychological autonomy of industrial trawler crews, in *Seafarer and Community: Towards a Social Understanding of Seafaring*, edited by P. Fricke (London: Crooms Helm), 47–64.

34 UK P&I Club 1999, *Analysis of Major Claims: Ten-year Trends in Maritime Risk* (London: Thomas Miller P&I Ltd).

35 The importance of allowing partners to sail has not gone unnoticed by some shipping companies. COSCO has recently begun to allow the wives of some senior officers to join their husbands aboard for a voyage in order to promote communication or 'mutual understanding' between the seafarers and their wives. As a senior manager explained, 'our intention is to provide opportunities for the wives to understand or appreciate what a hard job their husbands do'.

# 9 Seafarers of the world's largest fleet

## James McConville

## 1 Introduction

The seafarer is the archetypal international worker, employed on board vessels registered under differing flags, owned and operated by citizens of a multiplicity of countries. The international structure of the industry is further reflected in the numerous nationalities of the seafarers themselves. Such a diverse industry is dependent on international regulation to establish and ensure adherence to acceptable standards and conditions of employment.

Notwithstanding numerous conventions and procedures, substantial variations in living and working conditions between vessels operating under regimes of different flag states continue to exist, with particular concerns on the open registers. This study sets out to analyse the regulations and other criteria providing the qualitative framework underpinning seafarers' working lives on the Panamanian flag.[1]

The Panama economy is essentially based on service industries, which include offshore financial and legal provisions, the register of ships and, of course, the canal facility. The beneficial tax regime encourages such activities since only earnings generated in Panama itself are taxed. Similar to other South American countries, it experiences high levels of unemployment. In the year 2000, over 13 per cent of the working population were unemployed.

The maritime industry in Panama is perhaps unique. It constitutes the largest shipping register in the world responsible for some 100,000 seafarers. However, with the obvious exception of the canal, it lacks the basic elements essential to a maritime infrastructure. This is exemplified by the extremely small number of national seafarers who sail on their registered vessels. Recently the government has become increasingly conscious of the need for a maritime culture. To this end they have established the Panama Maritime Authority (PMA) with a remit to develop the necessary short- and long-term strategy for the country.

Intense international pressure and the international composition of the register have generated considerable debate relating to the protection of the seafaring labour force. The previous absence of adequate regulation in this area resulted in a presidential decree[2] specifically directed at seafarers working on board Panamanian flagged vessels.

This chapter explores the development of the register in terms of size, structure and ownership. It further discusses the need for a maritime culture and the importance of a maritime strategy. Finally, it highlights labour issues, both international and domestic, and examines implications for policy.

## 2 The shipping register

The analysis here considers the largest register of vessels in the world, sub-stantially larger than its nearest competitors. Table 9.1 illustrates the number of vessels, tonnage, and share of the total of open registers of the four largest of these flags in the year 2000. As can be seen, in terms of tonnage, the Panama register constitutes 40 per cent of the total and is double the size of its nearest rival, Liberia. In addition it is nearly three times larger than Liberia in numbers of vessels, the latter being the most important variable when considering seafaring employment.

Open registers can be defined as any flag state which does not require a substantial fleet for its own commercial purposes, but offers a legal base in return for fees from non-nationals to register tonnage which they own or control. There is the additional inducement of limiting the level of nation-alised regulation, both fiscal and operational, particularly the terms and conditions of employment of non-national seafarers. In the post-war period, the International Transport Workers Federation (ITF) consistently opposed them, defining open-registers as 'The beneficially owned ship and control of the vessel is found to lie elsewhere than the country in which the vessel is flagged'.[3] This loose definition caused a number of problems and was eventually dropped in favour of a list of countries provided what was termed the 'free-flagged facilities'. A UK enquiry[4] into shipping suggested a more generally accepted definition; this consisted of a number of ele-ments, which can be summarised as follows:

- the country of registry allows owners and those controlling merchant vessels to be non citizens;

*Table 9.1* Major open registers 2000 (1,000 GRT and over)

| | Number of vessels | DWT '000 | % share of total open registers (DWT) |
|---|---|---|---|
| Panama | 4,228 | 153.0 | 40.1 |
| Liberia | 1,420 | 73.1 | 19.2 |
| Malta | 1,341 | 42.9 | 11.2 |
| Bahamas | 1,011 | 42.8 | 11.2 |

Source: Institute of Shipping and Logistics July 2000.

- access to the registry and transfer from it is easy;
- tax on the income from ships was not levied locally or was low;
- the country of registry is a small power with no national requirements in any foreseeable circumstance for all the ships registered;
- manning on the ships by non-nationals is freely permitted;
- the country of registry has neither the power nor administrative machinery to effectively impose any government or international regulation, nor has the country the wish or the power to control the companies themselves.

As can be seen from the above criteria, the raison d'être of open registers is the creation of what has been termed 'regimes of immunity' as compared to traditional flags.

The Maritime Committee of the OECD commented that:

> The concept of the open register is the complete antithesis of the completely state-controlled transport of nationally generated cargo, and as such is philosophically very close the basic approach of Western countries. Unfortunately, the idea of the 'free flag' was for a considerable period associated with an abdication of the obligation that a state must effectively exercise its jurisdiction and control in administrative, technical and social matters over ships flying its flag.[5]

Another problem highlighted for traditional fleet operators related to the difficulties experienced by cross traders.

The most important element in the longer run, however, is the regrettable but probably inevitable decline of the role of the cross trader in world shipping. Almost by definition, ships on open registers are essentially operating as cross traders since, like traditional cross trading countries Norway and Greece (and indeed Hong Kong), the flag states are without substantial volumes of trade to and from their countries.[6]

Panama is the original modern open register which began in the years immediately following the First World War. As a phenomenon it was of little significance until the post-Second World War period when it began to accrue vessels from traditional registers. The early development of the register was due to the unique relationship between Panama and the United States which resulted in owners from that country transferring their tonnage to this new open register. In the latter half of the century, Greek ownership was also significant. While there had been a gradual expansion of numbers to 1970, the number and tonnage increased rapidly following the oil crisis of 1973. This growth was particularly vigorous over the last decade as can be seen in Table 9.2. Over the three decades, the numbers of vessels on the register have increased sixfold, but the tonnage has expanded to 16 times the 1970 level over the same period. This illustrates a trend towards larger vessels joining the register.

*Table 9.2* Panama Fleet 1970–2000 (100 GRT and over)

| Year | Number | Tonnage |
|------|--------|---------|
| 1970 | 886 | 5,645,877 |
| 1980 | 4,090 | 24,190,680 |
| 1990 | 4,828 | 42,911,700 |
| 1995 | 5,777 | 71,921,698 |
| 1996 | 6,105 | 82,130,668 |
| 1997 | 6,188 | 91,127,912 |
| 1998 | 6,143 | 98,222,372 |
| 1999 | 6,143 | 105,248,069 |
| 2000 | 6,184 | 114,382,270 |

Source: Lloyds Register Statistics in GRT.

It has been further estimated that between 1996 and 2001, the tonnage on the register increased on average by 10.7 per cent annually. A detailed table of fleet development is contained in Appendix 9.1.

Open registers are attractive because they afford considerable benefits to shipowners and operators, which result in a reduction in their costs. Any company irrespective of nationality may enter tonnage on the Panamanian Register. Assuming the vessels possess valid certification, there is no requirement for a survey prior to entry, the only exception being vessels over 20 years of age, which require an authorised inspection. Non-resident owners pay no taxation and there has been a commitment by the authorities that fees will remain stable. Discounts are also offered for substantial tonnage commitments. Furthermore dual registry facilities are also available. Under this system, a foreign vessel bare boat chartered for a period of two years can be registered for the same period without losing its previous registration, assuming that the original register gives appropriate authorisation.

In what follows there is an examination of the ships registered in Panama by type of vessel and country of ownership in order to identify contemporary trends. What emerges is the importance of bulk carriers and Asian ownership.

Table 9.3 highlights the growth in the register measured in both total numbers and tonnage over the five-year period 1995–2000. As can be seen, there is a strong concentration in general cargo and bulk trades both wet and dry, but there has been a change of emphasis. The proportion of general cargo vessels has contracted by number from 45 per cent to 36 per cent of the register but in terms of tonnage it has never been of importance, constituting less than 10 per cent. At the same time, the bulk trades, particularly dry bulk have become increasingly significant. In 2000, dry bulk carriers constituted almost 50 per cent of the register, which together with tankers represent more than 80 per cent of the total registered tonnage.

Table 9.4 concentrates on beneficial ownership, or more specifically,

*Table 9.3* Panamanian registered merchant fleet by major ship types (1,000 GRT and over)

| | Tankers | Bulk carriers | Container ships | General cargo ships | Passenger | Total |
|---|---|---|---|---|---|---|
| *Numbers* | | | | | | |
| 1995 | 713 (23.2%) | 695 (22.7%) | 224 (7.3%) | 1,368 (44.6%) | 68 (2.2%) | 3,068 (100%) |
| 2000 | 1,049 (22.7%) | 1,313 (28.3%) | 487 (10.5%) | 1,665 (35.9%) | 120 (2.6%) | 4,634 (100%) |
| *Tonnage (dwt)* | | | | | | |
| 1995 | 36,129 (42.0%) | 32,311 (37.5%) | 5,895 (6.9%) | 11,524 (13.4%) | 196 (0.2%) | 86,055 (100%) |
| 2000 | 50,321 (32.6%) | 75,461 (48.8%) | 13,984 (9.0%) | 14,380 (9.3%) | 401 (0.3%) | 154,547 (100%) |

Source: Institute of Shipping and Logistics.

*Table 9.4* Panamanian registered vessels by major country of domicile 1994–2000 (number and percentage) (1,000 GRT and over)

| Year | 1994 | 1995 | 1996 | 1997 | 1998 | 1999 | 2000 |
|---|---|---|---|---|---|---|---|
| *Japan* | | | | | | | |
| Number | 1,211 | 1,271 | 1,326 | 1,323 | 1,427 | 1,601 | 1,655 |
| % Share | 40.7 | 41.4 | 41.2 | 40.3 | 33.1 | 35.5 | 35.7 |
| Position | No. 1 | No. 1 | No. 1 | No. 1 | No. 1 | No. 1 | No. 1 |
| *Greece* | | | | | | | |
| Number | 311 | 325 | 339 | 335 | 447 | 503 | 523 |
| % Share | 10.4 | 10.6 | 10.5 | 10.2 | 10.4 | 11.1 | 11.3 |
| Position | No. 2 | No. 2 | No. 2 | No. 2 | No. 2 | No. 2 | No. 2 |
| *Hong Kong* | | | | | | | |
| Number | 268 | 254 | 261 | 245 | 240 | 250 | 235 |
| % Share | 9.0 | 8.3 | 8.1 | 7.5 | 5.6 | 5.5 | 5.1 |
| Position | No. 3 | No. 3 | No. 3 | No. 3 | No. 4 | No. 5 | No. 6 |
| *Korea* | | | | | | | |
| Number | 155 | 173 | 210 | 234 | 290 | 331 | 355 |
| % Share | 5.2 | 5.6 | 6.5 | 7.1 | 6.7 | 7.3 | 7.7 |
| Position | No. 4 | No. 5 | No. 4 | No. 4 | No. 3 | No. 3 | No. 3 |
| *Taiwan* | | | | | | | |
| Number | 152 | 186 | 199 | 201 | 232 | 252 | 277 |
| % Share | 5.1 | 6.1 | 6.2 | 6.1 | 5.4 | 5.6 | 6.0 |
| Position | No. 5 | No. 4 | No. 5 | No. 5 | No. 5 | No. 4 | No. 4 |
| *PR China* | | | | | | | |
| Number | 108 | 126 | 137 | 156 | 184 | 212 | 243 |
| % Share | 3.6 | 4.1 | 4.3 | 4.7 | 4.8 | 4.7 | 5.2 |
| Position | No. 6 | No. 6 | No. 6 | No. 6 | No. 6 | No. 6 | No. 5 |
| *USA* | | | | | | | |
| Number | 91 | 90 | 106 | 107 | 105 | 94 | 95 |
| % Share | 3.1 | 2.9 | 3.3 | 3.8 | 2.4 | 2.1 | 2.1 |
| Position | No. 7 | No. 7 | No. 7 | No. 7 | No. 7 | No. 7 | No. 7 |
| Total | 2,979 | 3,068 | 3,216 | 3,286 | 4,316 | 4,516 | 4,634 |

Source: ISL merchant fleet databases; aggregates based on quarterly updates from Lloyd's Registry of Shipping/LMIS.

country of owner domicile. The number of vessels is examined simply because this is the primary indicator of employment opportunities for seafarers. In the period 1994–2000, the number of vessels on the register increased steadily and significantly by a total of 1655 (58 per cent) vessels. This had a substantial impact on the number of seafarers required to man the fleet.

From the country of owner's domicile, it is clear that Panama is the primary Asian overseas register. In the year 2000, some 2,765 vessels (68 per cent) of the register were Asian owned. The distribution of country of domicile illustrates a number of interesting trends. The Japanese are undoubtedly the major owning country, and the number of their vessels has increased steadily throughout the period. Although the percentage share has seen a minor contraction, the proportion of Japanese owned vessels has remained at over one third of the fleet. The number of Greek owned vessels has grown consistently to maintain their 10–11 per cent share. Among the remaining Asian countries there has been some reordering of position with Hong Kong being the single example in the group where a contraction occurred. The US, despite its vital importance to the register's establishment and development, now owns under 100 vessels approximately 2.5 per cent of the total register over the period.

Being the largest register in the world, analysis of the fleet is vital to any discussion of the Panamanian industry and it implications for labour. The numbers of vessels make it the largest employer of seafarers in the world, and highlights the importance of a regulatory framework for their protection. The BIMCO/ISF Manpower Report in 1990[7] estimated that Panama employed a total of 103,000 seafarers, comprising 46,000 officers and 57,000 ratings. On a pro rata basis this implies employment in the year 2000 of 134,000. The ownership is largely in the hands of Asian domiciled companies, who also have a responsibility through contractual obligations to ensure adequate terms and conditions of employment.

## 3 Creating a maritime culture?

It has been suggested that Panama has all the resources to become a maritime nation but for historical reasons the opportunity has not been exploited. The major reason behind this wasted opportunity was the US control of the canal. During their period of ownership, the US was only interested in the strategic importance of the canal, not the wider maritime industry in Panama. Since its handover to Panama in December 1999, the canal has continued to function as an autonomous, albeit state-owned body, retaining its particular management and operational structure. It employs around 7,000 permanent staff and 2,000 casual workers. Of these, an estimated 2,500 are ex-seafarers who have been recruited with the basic training, and drawn from the national labour force.

On transfer back to Panama, the former US and other foreign canal workers were gradually replaced by Panamanians who were afforded the same benefits as the workers of a Federal State of the US. It was agreed that the Ministry of Labour and Labour Development would have no power over such terms and could not interfere in canal operations. The wide gulf in terms and conditions of employment is illustrated by the minimum wage of $1.22, which exists in the rest of Panama compared to that of $5 in the canal zone. Workers on the canal however, are not permitted to take industrial action since they are involved in the provisions of essential service.

The favourable conditions of employment created an eagerness to work for the Canal Authority, which soon reached full capacity, if not overcapacity, in terms of labour. This situation has been exacerbated by a reluctance on the part of many canal workers to retire at the official retirement age of 62. Since no further employment opportunities exist, seafarers currently training in the hopes of employment in the canal have little choice but to find opportunities elsewhere.

The attitude and culture of the Canal Authority is elitist and detached from the rest of Panama, let alone the rest of the maritime community. The canal is a separate entity, a business acting in isolation. Given the strategic importance of the canal to the maritime industry worldwide, the existence of the canal represents an opportunity on which to establish a maritime culture. Despite their aloofness, it is encouraging to note that the Canal Authority is involved in the formulation of a maritime strategy for Panama. The production of such a strategy is required by law under the jurisdiction of the recently established PMA. This organisation was conceived as part of the legislative decree of 1998[8] and has a number of functions under the decree (see Appendix 9.2).

The Director of the Maritime Authority, Mr Salazar realises the urgency of this maritime strategy and considers it to be necessary for the country and not just the government. Many institutions and sectors are participating in its construction, in particular the canal, organisations involved in tourism, environmental agencies, and universities. Mr Salazar emphasised the need for co-operation and the adequate preparation of the people of Panama particularly regarding education. In short, a maritime culture can be created whereby Panama produces an increasing number of seafarers and other maritime specialists necessary to service a wider maritime cluster. This will include shipping finance, shipbroking, marine insurance and the whole range of additional maritime services.

The hope is that the creation of such a maritime culture will improve the international perception of Panama as a genuine maritime nation, a perception which has been tainted by licensing scandals,[9] abandoned workers and non payment of wages. Such problems, including the lack of adequately controlled consular services are being addressed. As a result of corruption in the licensing department, 24 employees were dismissed, five of whom were senior officials. Security measures have also been tightened, with cameras installed at the offices combined with a number of other security features introduced to ensure that the Panamanian licences are less susceptible to fraud.

The Maritime Chamber is also a central player in the development of the maritime strategy. The Chamber has approximately 60 members representing ship agents, bunkering companies, banks, shipping lines, surveyors, shipyards, maritime lawyers, and railroads. The major and surprising omission from this list is shipowners. Panama, despite its large register, is home to no shipowners. This is an important issue for the development of the maritime culture and the chamber is eager to encourage owners into

Panama. The members of the chamber also recognise the need for comprehensive maritime data and detailed knowledge of how much the maritime industry adds to the economy.

Panama recognises the importance of the maritime industry and sees the opportunity and necessity to develop a maritime culture to support the register. To this end, the PMA has taken steps to establish a much-needed maritime strategy, which will create a genuine link between the register and maritime activities of the country.

## 4 Seafarers

The Decree Law No 8 of 1998 'Whereby work at sea and in navigable waterways is regulated, and other resources are dictated'[10] was drafted in recognition of the need to establish legislation and regulation specific to seafarers both national and international. The decree applies across the fleet regardless of nationality. It regulates recruitment, termination of contract, inspection and procedure regarding disputes. The decree consists of 12 chapters and 144 articles.

The decree is concerned with the 'relations between employers and workers on board ships engaged in international trade, domestic trade, the exploitation of living and non-living resources and other activities in the navigable waterway' (Article 1).

Under previous legislation 10 per cent of the crew on Panamanian registered vessels were required to be Panamanian seafarers. Although this stipulation no longer exists, the decree still promotes a preference for nationals.

> All shipowners of Panamanian vessels in international service shall try in the presence of equal conditions and capacity to give preference to Panamanian crew members and to foreign crew members with Panamanian spouses or with Panamanian children.
>
> (Article 4)

Further Article 15 states that manning agencies shall preferably hire Panamanian crew.

Apart from the positive discrimination in favour of the Panamanian seafarers the decree insists on equal treatment between nationalities. Article 5 states that seafarers should be paid the same salary for the same job performed by the same shipowner and it is forbidden to discriminate against anyone on account of union membership, religion, race or political affiliation.

The recruitment process is regulated by the decree, which requires manning agencies to be authorised by the Ministry of Labour and Labour Development. At least one of their agents should be a Panamanian residing in Panama. Such agencies may obtain fees from the shipowners but may not charge the seafarer for services rendered. Each manning agency is required to post a security bond with the Ministry of Labour of between $50,000 and $100,000.

Employment contracts must be signed by the shipowner or his representative and the crew member who is allowed to review the articles before he signs. A medical is compulsory before employment. On termination of employment, it is the duty of the shipowner to repatriate the seafarer either to the place of embarkation or place where contract was signed at the seafarer's request.

The details of hours of work, leave and overtime have been left to the employer, but overtime should be at least basic pay plus 25 per cent or time off in lieu of overtime.

Inspection procedures are the responsibility of the Panama Maritime. Decree No 8 itself states that inspectors should examine water and food supply, the kitchen and other facilities used to prepare and serve food, and certificates of competency of the vessel's catering staff. Crew members are within their rights to make a complaint to the PMA and special inspection is required in these cases. The PMA is also obliged to produce an annual report containing the information collected on accommodation and food and this should be made available to all interested organisations and individuals. The decree, in Articles 110 to 113, also covers the prevention of labour accidents, with regulations again laid down by the PMA.

Much controversy surrounded the implementation of the decree which was imposed by the President without the opportunity for discussion amongst the interested parties. The content was also in dispute, particularly Article 55, which allowed for dismissal without just cause. This was successfully challenged as being unconstitutional and has therefore been deleted from the legislation. Action against a further 14 articles are currently being resolved by the Supreme Court. The nature of such decrees is that despite a challenge, they are still considered law until the final decision is made.

Criticism has also come from the elimination of the 10 per cent rule relating to Panamanian seafarers and a series of other benefits such as leave pay, overtime and the absence of statements on collective bargaining. Although the decree law does not prevent collective bargaining, it does not oblige such agreements as did the Labour Code. This has been interpreted as implying that collective bargaining is only possible with the employers' consent. Consequently, many of the Panamanian unions are pressing for a return to the Labour Code for domestic workers. They also consider that too many of the issues relating to conditions of employment are not made explicit by the law and hence left wholly to the discretion of the employer.

Enforcement of the decree is perhaps the greatest problem. Despite articles stating that preference should be given to nationals, the unions in Panama claim that such positive discrimination does not occur and that there remains significant unemployment amongst their seafarers. One of the ratings' unions had 957 members of which only three were currently employed at sea. It was also clear from wider discussion that not all manning agencies were operating with the proper authorisation. Despite the implementation of these laws, enforcement is still a problem and this, as discussed earlier is one of the responsibilities of the PMA.

The major area of consensus amongst unions, and employers and the wider maritime community covered by the Maritime Chamber is that the decree is too broad and non prescriptive and that regulations are needed to define how the law should be applied. For this reason the Maritime Chamber together with unions have formed a working committee to draft such regulations. This initiative was welcomed by the Ministry of Labour and Labour Development.

In the domestic labour market,[11] seafarers are experiencing a number of difficulties. A major problem is the cost of training, which appears to be more expensive than in other similar countries. It was suggested that Panamanians who had undergone the necessary training were experiencing difficulties securing certificates. Officers have, in addition, gained a reputation among shipowners for limited sea employment in order to simply qualify for the work within the canal system. They are also considered expensive compared to Filipino, Croatian, Ukrainian seafarers. A further overarching problem was the quality of English spoken, which the nautical schools emphasised as a particular deterrent to their employment opportunities.

The Panamanian unions are therefore seeking tripartite discussion of decree law in order to create a more favourable environment with respect to employment opportunities and conditions.

## 5 Policy implications and conclusions

Panama was strongly persuaded by the US to establish the first maritime region of immunity through the introduction of the modern open register. The register provides a legitimate environment in which individual shipowning entrepreneurs can operate, but is unusual in that its considerable success has not engendered a maritime interest or culture within Panama. Thus, despite being the largest flag registry, there is a distinct lack of participation by Panamanians at all levels, but specifically seafarers. This is the serious challenge confronting the administration and in particular the recently established PMA.

The absence of maritime culture is perhaps surprising given the presence of one of the world's most important maritime arteries, namely the canal. The structure at the time of handover has ensured immunity to the canal area, which operates as a completely separate entity, divorced from the rest of the country. This attitude must be modified if Panama is to develop a comprehensive and cohesive maritime strategy. The canal together with the register provide a framework on which to construct a genuine maritime community. Recently, government and administration have realised the need to integrate these previously isolated and insular activities into the Panamanian economy and society as a whole. The effort to achieve this together with a drive to improve the perception and reputation of the register has been undoubtedly one of the most noticeable features of current government policy. The aim is to provide a forum in which economic and political differences can achieve some compromise in both the national and international interest. This of course implies initiating and extending national jurisdiction and providing the means of ensuring enforcement.

The decree of 1998 represented a serious attempt to address the fundamental gaps in maritime legislation and regulation. However, it is not without its own problems. The terms in which the decree is drafted leave much to the discretion of the shipowner's employer. Tripartite discussion on a strong regulatory regime to support it will assist in clarifying the broad issues and presumably lead to more effective enforcement.

What has been most pertinent to this analysis is that the government has recognised the importance, perhaps predominance, of maritime issues, both capital and specifically labour. Within the regime there is also a consciousness of the complexity of international shipping and its extreme sensitivity to political pressures and national and international regulatory attitudes. Panama must continue to seek to create a balance between competitive pressures and international respectability. It is hoped that shipowners will support these laudable aspirations.

## Appendix 9.1

*Table A9.1* The development of the Panama Fleet 1925–2000 (100 GRT and over)

| Year | Number | Tonnage | Average age | |
|------|--------|---------|-----|-----------|
| | | | Age | World age |
| 1925 | 18 | 97,566 | – | – |
| 1930 | 28 | 74,697 | – | – |
| 1935 | 42 | 136,859 | – | – |
| 1939 | 159 | 717,525 | – | – |
| 1950 | 573 | 3,361,339 | – | – |
| 1960 | 607 | 4,235,983 | – | – |
| 1970 | 886 | 5,645,877 | – | – |
| 1980 | 4,090 | 24,190,680 | – | – |
| 1985 | 5,512 | 40,674,201 | – | – |
| 1987 | 5,314 | 43,338,277 | 15 | 15 |
| 1988 | 5,283 | 45,369,271 | 16 | 17 |
| 1989 | 5,114 | 47,363,649 | 16 | 17 |
| 1990 | 4,828 | 42,911,700 | 16 | 17 |
| 1991 | 4,991 | 47,467,954 | 17 | 18 |
| 1992 | 5,424 | 52,485,614 | 17 | 17 |
| 1993 | 5,564 | 57,618,623 | 17 | 18 |
| 1994 | 5,799 | 64,170,219 | 18 | 18 |
| 1995 | 5,777 | 71,921,698 | 17 | 18 |
| 1996 | 6,105 | 82,130,668 | 18 | 19 |
| 1997 | 6,188 | 91,127,912 | 17 | 19 |
| 1998 | 6,143 | 98,222,372 | 16 | 19 |
| 1999 | 6,143 | 105,248,069 | 16 | 20 |
| 2000 | 6,148 | 114,282,270 | 16 | 20 |

Source: Lloyds Register Statistics.

## Appendix 9.2

### *Functions under the decree*

1  To propose, co-ordinate and carry out the National Maritime Strategy.
2  To recommend policies and actions, to carry out administrative actions, and to enforce laws and regulations referring to the Maritime Sector.
3  To orchestrate measures to safeguard the national interest in jurisdictional waters.
4  To manage, conserve, recover and exploit marine and coastal resources.
5  To co-ordinate with the Ministry of Agricultural Development in order to ensure that the country's aquaculture develops in strict compliance with the Panamanian State's international obligations, for which the Authority is primarily responsible.

6  To enforce the strict compliance of the provisions of the United Nations Convention on the Law of the Sea 1982 and other international treaties, conventions and instruments ratified by Panama regarding the Maritime Sector.

7  To evaluate and to propose to the Executive Branch and other government agencies that may so require, the adoption of those measures necessary in order to adopt international treaties and conventions relating to the activities carried out within the Maritime Sector.

8  To represent Panama before international organisations regarding matters pertaining to the Maritime Sector, in co-ordination with the Ministry of Foreign Affairs.

9  To co-ordinate with the National Maritime Service in order to enforce national laws in the maritime areas and internal waters of the Republic of Panama.

10  To update the signalling system, the navigation aids, the nautical charts and other hydrographic data needed for the safe passage of ships through the maritime spaces and internal waters of the Republic of Panama, in accordance with the provisions of the National Constitution and the laws of the Republic.

11  To direct, in co-ordination with other competent State agencies, the operations required to control oil and chemical spills, and any other disasters or casualties that may take place within the maritime areas and internal waters under Panamanian jurisdiction.

12  To co-ordinate with the National Institute for Renewable Resources, or its equivalent, the observance of the provisions of the United Nations Convention on the Law of the Sea, as well as national provisions, in matters relating to the protected marine coastal areas under the latter's responsibility.

13  Any other function assigned to it by Law.

## References and notes

1  Much of the present work is founded on Leggate, H.K. and McConville, J., 2002, Report on an ILO investigation into living and working conditions of seafarers: case study Panama (Geneva: International Labour Office).

2  The Decree Law No 8 of 1998 'Whereby work at sea and in navigable waterways is regulated, and other resources are dictated' (26 February 1998).

3  McConville, J., 1997, *Transport Regulation Matters* (London: Pinter), 86.

4  'Committee of Inquiry into Shipping Report, Chairman Lord Rochdale', London, May 1970, 51.

5  *OECD Maritime Transport*, 1981, 119.

6  Ibid., 121.

7  BIMCO/ISF 1990, Baltic and International Maritime Council/ISF, Manpower update.

8  Decree Law No 7 'Whereby the Panama Maritime Authority is created, the various maritime competencies of the public administration are unified, and other measures are dictated' (10 February 1998).

9 David Cockcroft, General Secretary of the ITF was able to buy a first mate's ticket certified by the AMP despite having no relevant qualifications or experience.

10 The Decree Law No 8 of 1998, Op. Cit.

11 N.B. The BIMCO/ISF Manpower Report 2000 states that there are 325 officers and 2,611 ratings. This appears to be an underestimation since there are a considerable number working in the canal area and one union also claimed to have over 2,000 affiliates. (There are three ratings unions and one officers union).

# Part III

# Ports in transition

*Edited by H. Meersman and E. Van de Voorde*

# 10 Ports as hubs in the logistics chain

*H. Meersman, E. Van de Voorde and T. Vanelslander*

## 1 Introduction

The maritime sector has, in recent decades, undergone two major 'revolutions' that have fundamentally altered the operational and organisational patterns within the sector.[1] First and foremost, increasingly large vessels were being deployed. In addition, there was a growing tendency towards the use of unit loads and containers.

One should, however, not lose sight of the fact that these revolutions in the maritime sector were directly or indirectly connected with important developments in the global economy and international trade. The trend towards globalisation remains a hotly debated topic, especially in the context of an international redistribution of labour and capital. Meanwhile, maritime transport has been growing spectacularly.

This rapid development in global trade and, consequently, in maritime transport, have also resulted in an altered role and significance of seaports. Modern seaports have become critical nodes in the complex network of logistical transport chains. Seaports that are unable to present themselves as key players in the optimisation process of the logistics chain to which they belong, or that are unfortunate enough to belong to a non-competitive chain, will become the victims of evolutions in the conception of international freight movement.[2]

The consequences of these significant developments for the port industry are clear to see. Shipping companies are looking to exploit every opportunity to realise benefits of scale, and they are deploying larger vessels and calling at fewer ports. At the same time, greater flexibility and speed are demanded from ports and port services. Shipping companies have, meanwhile, concentrated their efforts in all kinds of co-operation agreements, for example, the strategic alliances that have emerged in container traffic. This trend towards concentration has also led to a shift in the balance of negotiating power of the various port players. At the same time, shipowners are becoming increasingly interested in integrated transport and global logistics services.

All these developments have consequences for the goals and the organisation of ports, as well as for the future development of the port industry.

## 2 The purpose and organisation of ports

Meersman, Van de Voorde and Vanelslander[3] briefly explain the purpose of the organisation of a modern port. Besides the traditional basic activities of throughput of goods and passengers, new specialised activities have developed in the course of time, including forwarding and agencies. Moreover, maritime and hinterland modes were connected, so that storage came to occupy a central role (see Figure 10.1).

Thanks to various agglomeration effects (scale, location, urbanisation), some ports also became excellent locations for industry. These ports not only belonged to trade and transport chains, but they also developed into important links in industrial chains. Until quite recently the leading ports combine trade, transport and industrial functions. Under the influence of globalisation, they are increasingly developing into hubs from where the hinterland is supplied with imported products and from where goods destined for exportation are stored and shipped.

The purpose that ports serve in the distribution of goods is inextricably linked with the storage of goods. This is a consequence of the fact that the capacity of the hinterland transportation modes rarely corresponds to the amount of goods imported or exported by sea. Moreover, the unloading of a sea-going vessel need not necessarily happen concurrently with the loading of the hinterland mode. Furthermore, there are a number of derived activities that are not necessarily water-related, such as customs inspections and cargo preparation.

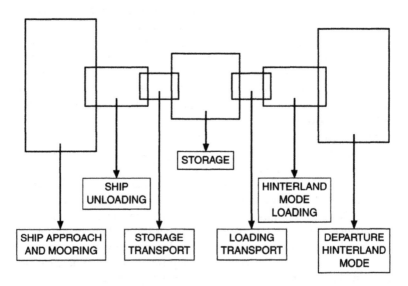

*Figure 10.1* Sub-processes of cargo throughput (source: Meersman, Van de Voorde and Vanelslander (2003).[3]

The diverse activities within the port creates its own internal links and at the same time the port itself is a link in a global logistics system. In the course of time, the relative significance of the various links within a port has clearly changed. This is partly due to efficiency-enhancing technological developments, including the increasing degree of containerisation, increasing ship sizes, quicker handling, etc.

This evolution in the functions of a port has made an impact on the organisational and managerial aspects of operations. After all, port activities involve a great many different parties at supervisory, managerial and operational level. These parties may be united in a single company, or they may constitute a myriad of enterprises and institutions within the port.

The port as a physical entity is run by a port authority, which in turn is usually partly or wholly supervised and/or regulated by a higher authority. Thus, government can be represented within a port to different degrees. Besides the port authority, any number of companies may have established themselves in a port.

A port authority can roughly subdivide the other market players into two groups, i.e. the port users and the service providers.[4] Among the port users are the shipping companies, the shippers and the industrial concerns that are located within the port perimeter and have acquired land in concession. The service providers are a heterogeneous group, consisting of pilots, towage, agents, forwarders, ship repairers, suppliers of foodstuffs and spare parts, waste reception facilities and bunkerers. Stevedores, who are evolving towards terminal operating companies, are a special case. They provide paid services (transhipment, storage, stripping and stuffing, etc.) to shipping companies and shippers, while they, in turn, pay the port authorities for a concession.

Port activities can also be approached from a different angle, i.e. by following the goods flow. The shipper calls on an agent and/or a forwarder to get his goods loaded onto the vessel of a shipping company. Shipping companies call on stevedores or terminal operators for transhipment and storage. Often, shipping companies call on the services of third parties (pilots, towage, agents, forwarders, ship repairers, suppliers of foodstuffs and spare parts, waste reception facilities and bunkering).

One thing is very clear though: within the port perimeter, there are a great many parties operating, each of which has its own objectives. Consequently, there is a considerable degree of heterogeneity, not only within ports but also between ports. Therefore, there is little sense in making general comparisons between ports. Indeed, any attempt to do so is further complicated by the fact that ports usually operate in very different economic, legal, social and fiscal environments. This heterogeneity is, moreover, not a static characteristic of ports, but is constantly and rapidly changing.

## 3  How shall the port environment evolve?

All market players operating within ports are constantly confronted with a number of uncertainties. What shall the future maritime market look like? What measures must the major maritime players (shipping companies, goods-handlers) take in order to maintain their competitive position? What measures will be imposed upon those same market players by the regulatory authorities? The future strategies of shipping companies and goods-handlers will be influenced not only by their own decision, but also by their response to decisions by others.

The development towards a far-reaching rationalisation which is already unfolding will undoubtedly continue in the foreseeable future. It will involve the deployment of new, larger vessels, which are expected to occupy a greater share of the market because of the lower operational cost per slot. Furthermore, it might involve new merger and co-operation agreements. In addition, there are landside technological improvements to take into account, as these will impact on the cost structure and the port environment. Finally, one may expect some shipping companies to redirect or diversify their activities.

The key question is where the trend towards larger vessel sizes will end. Wijnolst *et al.*[5] conducted a study to demonstrate that the Malacca-Max containership (18,000 TEU) is technically and economically feasible given the present state of technology. The main advantage of this vessel type is that it offers significant benefits of scale. According to the authors' calculations, this will result in a cost reduction by 16 per cent in comparison to the most cost-effective vessels that are available today. These cost savings will be realised mainly in fuel consumption and in Suez Canal dues.

In this perspective, a further scale increase, perhaps towards vessels that are twice the size of present ships, would appear self-evident. However, this reasoning takes into account direct vessel costs only, and it largely ignores terminal costs and costs associated with hinterland transportation. Yet, the latter costs have proven before to be crucial to the choice of vessel type and port.

Moreover, there is a further industrial-economic issue to take into account. These types of super vessels will, for physical as well as economic reasons, only call at a limited number of ports. Shipping companies will, however, want to make sure that they cannot be held hostage by a small number of ports of call. With the example of the shipping of crude oil by means of ULCCs still fresh in the memory (i.e. the dependency of these vessels on the port of Rotterdam for the supplying of Europe), they will realise all too well that a lack of alternatives can inspire port authorities to increase prices.

In fact, the deployment of new and increasingly large vessels can be seen as an attempt by the large shipping companies to continue to put pressure on their competitors. In this manner, they aim to further reduce the cost per slot. The overcapacity that this creates is simply accepted as an unfortunate side-effect.

Some shipping companies will try to follow this strategy. They will also aim at a larger operational dimension in order to reduce the cost per slot. This might be achieved through takeovers or mergers, as other shipping companies are confronted with a similar market environment. Mergers and/or takeovers also enable companies to implement further savings plans and, if need be, to reduce capacity.

This pressure to rationalise will, in the short to medium term, result in a reorganisation of services, and new alliances will emerge. In the first instance, these alliances will seem as fragile as those of the recent past, because they are subject to strategic moves by shareholders and the management of the controlling holdings. Stability will only be achieved at the moment when the alliances are partly or wholly transformed into mergers and takeovers. At that moment, there will be fewer players in the market and it will have become easier to stabilise prices through, for example, capacity restrictions.

What developments may be expected at landside? The economic benefits that shipping companies seek in scale increases and associated cost savings should not be allowed to be wasted because of landside bottlenecks. The evolution towards a far-reaching automation will persist, as a result of, among other things, shortages of skilled workers, the high cost of labour and demand from customers for a 24/7 service.

The question arises, however, what will be the consequences of such a far-reaching co-operative agreements and takeovers in terms of the goods handling? To what extent will takeovers imply that companies attempt to impose technologies that are used in their home port upon terminals which they control in other ports or continents?

In addition to using larger vessels, and entering into mergers or co-operation agreements, some shipping companies may opt for other rationalisation strategies. As in other sectors of the economy, a tendency may emerge within the shipping industry to diversify into complementary activities. In this manner, a company becomes less vulnerable to economic trends within a single sector. Moreover, in the long term, such a strategy may tie in with a striving towards vertical integration, whereby control is acquired over a large proportion of the logistics chain as possible. Ultimately, this is also the goal of a number of shipping companies who present themselves on the market as integrated logistics service providers.

Thus, the international maritime chain has in recent years become increasingly integrated, often thanks to the initiative taken by shipping companies. The consequence of this evolution is a significantly greater market power for the large shipping companies *vis-à-vis* other service providers, including port authorities and goods-handlers.

The other market players in the logistics chain were confronted with a choice: either to undergo this evolution and continue to be dominated or even taken over, or, to respond quickly and effectively through similar initiatives. The latter is what actually happened. Port authorities launched new initiatives; and within the goods-handling sector, mergers, takeovers, and regional and global expansion drives changed the structure of the old stevedoring business quite fundamentally.

These structural changes have heralded a new phase in the process of port competition. Henceforth, the large shipping companies no longer face relatively small and insignificant players in negotiations over rates and volumes; instead, they are confronted with a limited number of large terminal operating companies who operate in different ports and can thus negotiate over larger packages. This implies, first and foremost, that port competition is no longer present at the level of port authorities, but at the level of private terminal operators, who are currently offering regional networks that will eventually become global networks.[6]

In this manner, the negotiating power of the shipping companies will, in the first instance, be undermined. At the same time, however, they will become increasingly interested in acquiring a dedicated terminal (perhaps in collaboration with a terminal operating company) or in taking a stake in a terminal. This evolution will reduce the footloose nature of the shipping companies. After all, acquiring a financial stake in a terminal operating company will encourage shipping companies to develop longer-term relationships and stop them from diverting to other, competing, ports.

The question arises, what will remain of the present port authorities? In the current 'game' of negotiations between shipping companies and terminal operating companies, they have the advantage of being able to allocate concessions. Eventually, a number of port authorities will no doubt feel the urge to use these concessions as a lever towards a more active role in the competition and throughput process. In the longer term, this may give rise to a neutrality issue, i.e. in relation to the allocation of new concessions.

From the overview above, we may conclude the following. In the short to medium term, major changes will occur in the fields of port competition and terminal management. These developments will bring with them a considerable degree of uncertainty, which will only be resolved if a number of questions can be answered adequately. Insight is required, for example, into the manner in which a further concentration of terminal operating companies and the active participation of port authorities will affect price setting in port services and influence the development of a competitive dynamics.

## 4 Selected works

The maritime and port sectors are typically dynamic sectors that are constantly subject to exogenous and endogenous influences. Consequently, the port environment has a tendency to evolve very quickly. It is therefore self-evident that the issue of port competition is such a hot topic.

Recent research into port competitiveness[7] demonstrates quite clearly how port competition must be approached dynamically from a scientific perspective. Thus, there is a need to constantly develop new theories and models, so that future evolutions and possibilities could be assessed as adequately as possible. Knowledge about the future begins with a careful analysis of the present.

The contributions included in this section constitute an excellent starting point for fresh and indispensable research. Each chapter deals with an aspect of port economics, invariably within the context of port competition and competitiveness.

The contribution by Heaver, Meersman and Van de Voorde (see Chapter 11) focuses on the response of port authorities to the changing market environment in which they operate. It documents the changes taking place in the relationships between port authorities and terminal management companies and considers the strategic issues faced by these groups and other port interests. In particular, it investigates the potential conflicts of interest for a port authority in matters related to the level of competition among terminals within a port and the amount of competition among ports.

J. Blomme's contribution (see Chapter 12) also takes into account the trend towards concentration. He notes that a number of distortions have been created in terms of market power. As a result, a number of smaller market players will either disappear or they will be forced to adopt a niche strategy in order to survive. Port authorities have a lot to gain from retaining a degree of internal competition, e.g. in stevedoring. Dedicated terminals need not be rejected on principle, because they can help tie a goods flow to the port. Nevertheless, the allocation of dedicated terminals must happen in accordance with special conditions. Co-operation and networking among port authorities are still in their infancy.

The continuing growth of container traffic and the increasing ship sizes puts strong pressure on ports and port terminals. In some load centres diseconomies of scale may result from lack of space and port congestion. T. Notteboom (see Chapter 13) explores some questions related to the development of container port systems and in particular the rise of new terminals and former non-hub terminals at the expense of existing large load centres. The central hypothesis put forward is that the shift of cargo from large ports to smaller or new ports, called the peripheral port challenge, is an inherent ingredient of contemporary port system development. The West Mediterranean port system is shown to be a prime example of a peripheral port challenge triggered by the need to decrease diversion distances.

## References and notes

1 Stopford, M., 2000, *Maritime Economics*, 2nd edn (London and New York: Routledge).
2 Van de Voorde, E. and Winkelmans, W., 2002, A general introduction to port competition and management, M. Huybrechts, H. Meersman, E. Van de Voorde, E. Van Hooydonk, A. Verbeke, W. Winkelmans (eds) *Port Competitiveness. An Economic and Legal Analysis of the Factors Determining the Competitiveness of Seaports* (Antwerp: Editions De Boeck Ltd), 1.
3 Meersman, H., Van de Voorde, E. and Vanelslander, T., 2003, *The Industrial-economic Structure of the Port and Maritime Sector: an Attempt to Quantification* (Palermo: NAV'2003).
4 Meersman, H., Van de Voorde, E. and Vanelslander, T., 2003, *The Industrial-economic Structure of the Port and Maritime Sector: an Attempt to Quantification* (Palermo: NAV'2003).
5 Ibid.
6 Notteboom, T., 2002, Consolidation and contestability in the European container handling industry, *Maritime Policy & Management*, 29, 268.
7 Huybrechts, M., Meersman, H., Van de Voorde, E., Van Hooydonk, E., Verbeke, A. and Winkelmans, W. (eds), *Port Competitiveness. An Economic and Legal Analysis of the Factors Determining the Competitiveness of Seaports* (Antwerp: Editions De Boeck Ltd).

# 11 Co-operation and competition in international container transport

## Strategies for ports

*T.D. Heaver, H. Meersman and*
*E. Van de Voorde*

## 1 Introduction

Increased horizontal and vertical integration is occurring amongst organizations involved in the international logistics of manufactured goods. The development of supply-chain management practices is evidence of the greater integration in the structure and practices of shippers. The global alliances among shipping lines, the growth of container terminal management companies with global operations and the increase of logistical service offerings by transport companies are evidence of the increased integration amongst suppliers of logistics services.

The level of integration is such that the demarcation between previously separate markets for logistics services is now blurred. For example, the value of treating shipping as a separate market has been eroded by the expansion of shipping companies into port-terminal management and into logistics service offerings. At the same time, organizations with expertise in container terminal management have been able to enlarge their roles in logistics services by managing terminals in different ports and by participating in the integration of ports with inland transport services. These and related developments in market structure within which international shipping and other logistics services are provided raise serious strategic questions for port authorities.

This chapter focuses on the response of port authorities to the changing market environment in which they operate. It documents the changes taking place in the relationships between port authorities and terminal management companies and considers the strategic issues faced by these groups and other port interests. In particular, it investigates the potential conflicts of interest for a port authority in matters related to the level of competition amongst terminals within a port and the amount of competition amongst ports.

The role played by individual organizations in international logistics systems as vertical and horizontal re-organization proceeds is uncertain. It is influenced by many factors, especially those affecting the power of organizations and their potential contributions to efficiency and profitability of logistics systems. In the rivalry to establish roles that meet the objectives of the various organizations (and, hopefully, the needs of shippers), the asymmetry in their positions is important. Two aspects of an organization affect its power. The first is its financial strength which is influenced by many factors. It is a key to an organization's ability to follow costly strategies to achieve its objectives. The second factor is the range of options available to it. This is dependent on the characteristics of the market structure. Disparities in the number of options available to negotiating parties can result in an imbalance of power. For example, a port authority without alternative locations is at a disadvantage in relation to a shipping line with a choice of ports. However, a port in a prime location derives an advantage from that in negotiating with alternate shipping lines.

The rivalry amongst the various players in the development of international transport and logistics systems involves the working of complex forces in the evolution of the new vertical and horizontal relationships. The outcomes affect the public interest in the preservation of structures and practices consistent with effective competition. Whilst the power of port authorities appears to have weakened, at least temporarily, they still play an important role in determining the development of the new systems.

## 2  Co-operation agreements between market players

In a previous study of the structure and scope of co-operation agreements in the maritime sector in the broad sense, one considered the effects of mergers and alliances on international shipping and port competition.[1] It was concluded that shipping companies, consortia and alliances have acquired a more powerful negotiating position *vis-à-vis* port authorities, terminal management companies[2] and inland transport firms. In general terms, the situation has developed towards three major blocks:

- a number of powerful alliances/shipping groups (the Grand Alliance, the New World Alliance, and the Global Alliance);
- a number of co-operation agreements regarding slot exchange; and
- a number of 'soloists': MSC (Mediterranean Shipping Company), Maersk Sea-Land (controls Safmarine), Evergreen (controls Lloyd Trietino and Unigbory), Cosco, China Shipping Group.

The situation is still evolving, with the vertical concentration in liner shipping still continuing. Alliances are starting to opt for a further concentration on certain terminals, with important consequences. Alliances and other co-operation agreements are now controlling significant goods flows on the major routes on which they are deploying larger vessels. This creates demand for ever-greater terminal capacity, where handling is organized in such a way that turnaround times are limited to 12 hours. Ties with inland logistics can give considerable control over the logistics chain.

At the same time, the other parties have responded to the changing challenges and opportunities by themselves entering into various forms of co-operation (strategic alliances, mergers, etc.) to achieve greater control of the logistics chain. These developments bring with them a danger of preferential treatment, conflict of interests and market dominance (*Maritime Policy & Management*,[1] p. 369).

By way of illustration, Table 11.1 provides an overview of the principal forms of co-operation between stevedores and port authorities that have materialized in recent years. These developments are resulting in continuous shifts in the balance of the markets. In order to gain a better understanding of this development, one puts forward the following pertinent research questions:

- How can one explain from an industrial and economic perspective the observed co-operation trends involving port authorities and container terminal management companies (CTMCs), the principal port players? More specifically, one is interested in the objectives, the means used and likely impact of these developments.
- What are the potential consequences for the parties involved?
- Which other trends can be discerned?

First and foremost, answering these questions requires an in-depth knowledge of the relevant environmental factors and a detailed analysis of the various forms of co-operation.

## 3 Developments within ports: the relevance of ownership and scale

The actions of port authorities are ultimately derived from their basic objectives, which are influenced by their ownership, structure and mandate. These matters are not the focus of this paper, but some observations about them are needed before discussing the structural changes in transport markets.

*Table 11.1* Co-operation agreements between various market players

| Market players | Shipping companies | CTMCs | Hinterland transport | Port authorities |
|---|---|---|---|---|
| CTMCs | Financial stake of shipping company in CTM; joint ventures; dedicated terminals | Participation in capital | Joint ventures | Financial stakes port authorities |
| Port authorities | Dedicated terminals | Financial stakes port authorities | Financial stake in hinterland terminals | Alliances between port authorities |

Source: Based on Heaver *et al.*, 2000, *Maritime Policy & Management*,[1] p. 365.

Increased private participation is found in ports of most countries.[3] However, full privatization of previously public ports is rare, mainly confined to the UK. Continued public presence in most countries in one way or another may leave the objectives of port authorities uncertain. Should they consider their primary objective to be maximizing the tonnage handled? Should they maximize the value-added activities within the port perimeter? Or, should they, perhaps, maximize profit-generating opportunities for industry and services located in the port?[4] Should the community provide funds to support port investments or should port investments provide a revenue source to the local community?

Differences in the objectives to be pursued by port authorities will affect their policies with respect to the amount and nature of competition they would expect among terminals within a port and amongst ports. A revenue-seeking port (for its local community or a dividend-seeking higher level of government) may not favour inter-terminal competition if total traffic is inelastic to price. It may, then, prefer limited competition and economic rent-seeking rates.

Competition as an issue for port authorities is influenced by the geography of port jurisdiction. The distance between ports and their potential to serve a common hinterland affect the levels of co-operation and competition. These relationships are not static, as they are affected by ocean, port and inland costs, but they tend to persist. Thus, the geography of Australia is less conducive to inter-port competition than that of Western Europe or North America. The jurisdiction of authorities affects the potential for competition; for example, in the US, county and city authorities give rise to the competitive adjacent authorities of Seattle and Tacoma and of Los Angeles and Long Beach, respectively. In other places, for example Georgia, the state is the authority, and the Port Authority of New York/New Jersey is a by-state authority. However, while local economic and institutional factors affect the potential for co-operative and competitive arrangements, the influence of developments in transport and logistics tend to be general.

In recent years, a number of structural changes have occurred in the transport market as firms have attempted to be price competitive and to improve and expand service offerings. This development is a result of striving towards cost-savings through expansion and the emergence of the concepts of global and total logistics. As a result, the principal players (e.g. shipping companies, terminal operators, forwarders) evolved into large logistical organizations through a mixture of autonomous growth, alliances, mergers, etc. Each of these organizations separately strives explicitly towards expanding its sphere of influence, as is evident from the example of shipping companies who, beside their primary transport activities, are increasingly involved in operating terminals and hinterland transport modes.

Public and private port authorities, on the other hand, with a few notable exceptions, may be seen as responding re-actively to the organizational developments about them. Most have responded, some more actively and effectively than others, to the logistical needs of shippers and of their immediate customers, the shipping lines, by initiatives in such fields as facility expansion and information technology. Public port authorities, especially, tied to their local jurisdiction, have been faced with the need to respond to the growth of container traffic and the increased power of fewer players in the logistics chains. The most notable example of a port that has gone beyond this is the Port of Singapore Authority, which has been pro-active in the marketing of its IT technology internationally and in entering into terminal/port management agreements internationally. CTMCs, on the other hand, for example P&O Ports and Hutchison Port Holdings (HPH), have been active in expanding the geographic and service scope of their businesses.

Tables 11.2 and 11.3 illustrate the shifts that have taken place in the

*Table 11.2* Container traffic in Antwerp – share in general cargo traffic (1975–2002)

| Year | Container traffic (TEU) | Container traffic (tons) | General cargo (tons) | Containerization rate |
|------|------------------------|--------------------------|----------------------|-----------------------|
| 1975 | 356,194 | 3,335,558 | 22,055,164 | 15% |
| 1980 | 724,247 | 6,125,967 | 28,459,417 | 22% |
| 1985 | 1,243,009 | 10,921,320 | 37,601,995 | 29% |
| 1990 | 1,549,113 | 16,553,429 | 43,522,594 | 38% |
| 1995 | 2,329,135 | 25,795,560 | 50,674,572 | 51% |
| 1996 | 2,653,909 | 29,460,184 | 52,206,146 | 56% |
| 1997 | 2,969,189 | 33,426,642 | 56,443,291 | 59% |
| 1998 | 3,265,750 | 35,376,283 | 60,150,697 | 59% |
| 1999 | 3,614,246 | 39,442,240 | 60,299,056 | 65% |
| 2000 | 4,082,334 | 44,525,643 | 68,737,166 | 65% |
| 2001 | 4,218,176 | 46,409,921 | 68,334,635 | 68% |
| 2002 | 4,777,151 | 53,016,582 | 73,336,386 | 72% |

Source: Antwerp Port Authority.

*Table 11.3* Concentration of container traffic in Antwerp (TEU)

| Market player | 1985 | 1990 | 1995 | 1999 | 2002 |
|---------------|------|------|------|------|------|
| Hessenatie (HN) | 347,419 | 580,033 | 1,094,921 | 2,297,246 | 3,966,667[1] |
| Noordnatie (NN) | 153,033 | 282,072 | 419,928 | 742,754 | |
| Total for the port of Antwerp | 1,243,009 | 1,549,113 | 2,329,135 | 3,614,246 | 4,777,151 |
| Share of HN + NN | 40.3% | 55.7% | 65.0% | 84.1% | 83% |

Source: Antwerp Port Authority and various stevedores

Note
1 extrapolated from April–December figures of PSA Antwerp.

relative influence of a number of players. Table 11.2 shows the spectacular growth in container transshipment in the port of Antwerp since 1975. Container traffic has grown through the increased containerization rate, i.e. the relative share of container traffic in overall transshipment, and through the growth of trade.

Table 11.3 shows that the two main stevedoring companies in the port of Antwerp have each increased their share of the traffic, and to a total of 84 per cent. This is mainly due to Hessenatie, which saw its market share expand from 27.8 per cent in 1985 to 63.6 per cent in 1999. The concentration trend is even more outspoken in the port of Rotterdam, where ECT achieved a market share of 70.9 per cent in 1999. A similar trend can be observed in Germany, in part because of the creation of Eurogate through a merger of the container activities of Eurokai in Hamburg and Bremerlagerhaus Gesellschaft in Bremerhafen.

A comparable development has taken place among shipping companies, where expansion and concentration trends have been apparent in relation to the control of slot capacity in container traffic. Whilst, in 1995, the 15 largest global shipping companies control 62.5 per cent of slot capacity, by 1999 their share had increased to 66.2 per cent.

The typical port authority has few opportunities to participate in developments beyond its jurisdiction. Take the port authority of Antwerp, for example: it concentrates on such activities as the deepening of the river Scheldt in order to retain its status of port of call for so-called mega carriers on the East-West.

In addition, the 'local' port authorities are increasingly confronted with large, international goods-handlers (e.g. P&O Ports, HPH, PSA Ports) whose international asset base, IT systems and cash flows are relevant to their efficiency, attractiveness and clout in negotiations. A typical example is PSA, which in 1999 embarked on three foreign projects, including Sines Container Terminal in Portugal. This brought the number of ports and terminals developed and/or operated by PSA at that time to ten. This is where the question arises of possible abuse of monopoly power. To what extent can a market player that controls several ports or terminals devise global strategies for shipping in conjunction with mega carriers, as a result of which other terminals and ports, that are not controlled by this player, may lose traffic?

The threat of excessive market power by dominant suppliers of port services is influenced by the size of ports and the number of their terminals, the level of competition amongst ports and the potential for new ports and new logistics routings. It is striking how new hub and terminal ports continue to enter the market.

## 4 Market players in ports: positions and strategies

There are three types of goods-handlers active in the container business. They are port authorities, shipping lines with terminal operations, and independent CTMCs. The role of these groups is determined in part by the policies of port authorities and in part by broader public policies as they may affect the public interest in achieving efficiency through open and competitive regimes. The following text examines the positions of the parties and the strategies that they may follow.

### *4.1 Port authorities: scope for action*

The discussion here of strategies for port authorities is confined to two areas. The first is the strategy of authorities with respect to the granting of terminal concessions. Port authorities face two primary issues in their decisions about terminal concessions. The first is who should operate a terminal and what are the conditions under which operation will be allowed? The second is the strategy of ports with regard to the competitive relationship amongst ports.

Port authorities are compelled to adapt their strategies in light of the abovementioned transformations in the structure and practices of other participants in international logistics. The response of shipping lines and others to the needs of shippers for improved logistics services are leading to a variety of pressures on ports.

#### *4.1.1 Issues with terminal concessions*

Ports are confronted with increasingly influential immediate customers. Shipping lines not only formulate demands with regard to port charges but they are more widely interested in the use of dedicated terminals. Dedicated terminals have been common in the US, where integrated systems to facilitate intermodal transport have been a priority. In Europe and elsewhere, as container volumes enable better utilization of such terminals and as the benefits of integration through corporate responsibility for planning, investment and operations management increase, so the interest of lines in dedicated terminals has increased. Shipping companies see the terminals as part of their international networks of transport and logistics services. Recent developments in Rotterdam are quite telling. Maersk/Sea-Land now has its own dedicated terminal; P&O Nedlloyd and the other members of the Grand Alliance have also been granted a dedicated terminal; and the World Alliance has moved to ECT's Delta Dedicated West Terminal. The market share of the large multi-user terminal provider had diminished. The development poses new issues for the smaller shipping companies, although there is no public evidence from American experience of discrimination against such lines. The interest of lines in terminal management also has implications for the market share available to CTMCs.

For port authorities, dedicated terminals are a means to facilitate the development of integrated services and to bind shipping companies to terminals. They provide opportunities for port authorities to push for more investment and longer-term leases than might otherwise be possible. The power to dedicate terminals is a useful strategy for a port authority, certainly if there is competition between different terminal operators. It is, also, a legitimate question whether European port authorities should/ could/may make more effective use of this current position of power, for example, by forming alliances (including through participation, such as the port of Rotterdam's stake in ECT) or by establishing networks as done by PSA. It also remains to be seen whether port authorities are capable of establishing the same kind of functional relationship with inland links in the logistics chain.

The choice of terminal operator may affect the amount of inter-terminal competition in a port. For a small port needing a second terminal, should a second operator be selected for a new terminal? Where a third terminal is to be added, are the economies of scale and scope such that one of the incumbents should get the additional terminal? What influence should mergers in the terminal operating business have on port authorities' policies about inter-terminal competition?

Little evidence is available on the working of competition amongst terminals. As terminals have shifted to integrated elements of the logistics systems of which they are a part, so the importance of system competition has increased over that amongst the elements. Therefore, the attention of port authorities to retaining competition needs to be greater when there are limited port routing options. Competition amongst terminals can be expected to be more of an issue in Australia than it is in North America or Western Europe.

Port authorities need to consider the effects on efficiency of their agreements with terminal operators. Long-term leases encourage optimal development strategies by terminals. Port authorities and governments that may play a roll in setting payment terms need to consider the effects of alternatives on efficiency. Direct charges to traffic, as common in Australia, or dividend rates variable with profitability, may be attractive to public bodies interested in revenue enhancement but they are not consistent with maximizing the efficient development of traffic. The use of long-term concessions to a few companies is also leading to some new investment strategies by port authorities, as found in Rotterdam and Antwerp.

In 1999, the Port Authority of Rotterdam acquired a 35 per cent financial stake in the major stevedore ECT, a company that in turn was controlled by HPH. The Port Authority also granted a joint dedicated terminal to ECT (a 33 per cent stake) and Maersk, while another was allocated to P&ONL in a 50/50 joint-venture with ECT. This inevitably raises questions. The move meant that the port authority of Rotterdam was investing in the risk-bearing capital of a private terminal operator. This could be considered as a typical example of a private-public partnership but, at the same time, it could be seen to compromise the impartiality of the port authority. After all, through its stake in ECT and the joint-venture, the port authority also acquired an interest in the wellbeing of Maersk Sea-Land and P&ONL.

A similar situation could occur in Antwerp. When the concessions for two new container terminals were granted (June 2000) to, respectively, a British consortium (P&O Ports, P&ONL and the port of Duisburg) and a Belgium combination (Hessenatie – Noordnatie – NMBS), the port authority imposed a number of conditions. Besides conditions concerning traffic, the port authority also wanted a number of clauses to be incorporated which give it the right to acquire a minority stake in the two consortiums. More specifically, it wants right of pre-emption if one or more of the groups involved decides to sell their own shares. Theoretically, this could enable the Antwerp port authority to acquire a stake in the port of Flushing via a sharehold in Hessenatie.

These examples are illustrative of two developments. Port authorities have recently started moving to acquire a more active position in the marketplace and the logistics chain. The power to grant concessions for container terminals to CTMCs or consortiums of them represents a useful tool in this respect, as it could enable port authorities to realize their goal in the short-term. At the same time, though, this evolution poses a threat to the impartiality of post authorities *vis-à-vis* other players (including competing terminal management companies and shipping companies), which will inevitably have repercussions for its role as a regulator in a port.

### 4.1.2 Issues of inter-port competition

Ports need to consider the status of competition from other ports and the logistics systems through them. The effects of improved intermodal services are to increase the competitiveness of alternate port routings. In addition, new ports can emerge, for example Amsterdam and Flushing in the Hamburg-Le Havre range. Initially, such new competitors may not pose much of a threat, but some gain a critical mass of traffic and establish effective hinterland connections. Monitoring the effectiveness of new ports requires careful attention to the success of their network strategies, even at the level of agencies and forwarding firms.

The interests of port authorities and of companies operating terminals in co-operation and competition with other similar enterprises are subject to similar considerations as relationships between other competitive businesses. The economies of scale and scope may create efficiencies at least to some level of co-operation, but the mitigating effects on competition must be taken into account. For small ports, even a merger may be beneficial when sufficient competition from other facilities remains. For example, the merging of the ports of Copenhagen, Denmark, and Malmö, Sweden, enabled by the Oresund Fixed Link, may be viewed in this way. Merging or price agreements between larger ports would be more likely viewed as anti-competitive.

Ports and terminals in close proximity often enter into small co-operation agreements with neighbouring facilities. For example, in Europe, the ports of Antwerp, Ghent Terneuzen and Flushing are considering more 'structural co-ordination'. The port authority of Rotterdam, for its part, has reached a co-operation agreement with Flushing, including with regard to joint investments. In the US, the Northwest Terminal Operators Association facilitates collective general marketing and lobbying on matters of mutual interest. The Association does not discuss pricing. In the US, ports enjoyed the right to exchange price information under the Shipping Act in the same way as conference lines, but discussions and agreements on pricing did not take place. This may be explained, in part, by the ownership of ports by local authorities more interested in the promotion of traffic through their port than in the maximization of profits.

At meetings of port authorities in venues such as the Association of American Port and Harbour Authorities, speakers have sometimes suggested that co-operation amongst port authorities might be the answer to the 'unreasonable demands' from shipping companies and alliances for more space and deeper-draught ports. However, consideration of such action or exchange of information about prices would likely raise serious issues under competition policies. In Europe, an intervention by the European Commission is looming, as shipping companies wish to retain the freedom to negotiate with ports separately, rather than with a single representative body.

### 4.2 Terminal operators: mergers on the horizon?

The restructuring of international shipping and logistics systems is also putting pressure on the companies providing container handling and terminal management services. The pressures on these businesses to provide high quality local service levels at very competitive prices while fitting into the global requirements of large lines and shipping alliances are forcing change in the industry. Three patterns of response are evident. They are mergers that result in immediate expansion and a stronger negotiating position, regional coverage that creates greater flexibility in supplying services in neighbouring ports and continuing globalization. Each of these warrants attention.

#### 4.2.1 Mergers

Faced with mergers and alliances amongst their customers, the development of global CTMCs and the need for ever larger investments to meet transshipment requirements through mega terminals, stevedoring companies are merging. An example is in Belgium. In June 2000, on the occasion of the granting of a concession for the third phase of the new Deurganck Dock on the left bank of the Scheldt in Antwerp, the two largest stevedores in the port, Hessenatie (a subsidiary of CMB) and Noordnatie decided to proceed with a merger. This created one of the biggest container transshipment companies in Western Europe, with a market share in Antwerp of almost 90 per cent. This merger created the opportunity to capitalize on synergies (economy of scale effects) and to attain a stronger negotiating position *vis-à-vis* shipping companies and the port authority (*Economic of Strategy*,[5] p. 98). Thus, the companies are able to take full advantage of the trend towards integrated services at increasingly competitive prices. Other pragmatic considerations also came into play. For example, the merger precludes a difficult choice of terminal for the Grand Alliance and the Americana shipping company. As an integral part of CP-Ships, Americana had previously had a contract with Hessenatie (which, due to a shortage of capacity, was actually executed by Noordnatie), while the Grand Alliance collaborated with Noordnatie. It speaks for itself that any joint operation by Grand Alliance and Americana would involve only one terminal.

However, mergers often have side consequences. In this case, the Belgian national railway company, which had until then been involved with Noordnatie in a joint venture at another terminal, became an interested party. Further, while the focus of attention is on container operations, it also affects the handling of some 45 million tons of general cargo. A consequence of the merger is to place the merged company in a dominant position in Antwerp in the short run.

### 4.2.2 Regional coverage

Another way for stevedoring companies to serve markets more effectively and to gain market power is to increase the intensity of their operations in a region. This may be done by increasing the span of services offered, a common trend in logistics-service companies, and by providing similar services in adjacent locations. The Antwerp-based goods handler Hessenatie, for example, has also established itself in the port of Zeebrugge for reasons connected with operational and commercial flexibility. As shipping companies are constantly moving from Antwerp to Zeebrugge and vice versa, a presence in both ports reduces the risk of client-loss for the company.

### 4.2.3 Internationalization

The increased commercialization of ports and the global expansion of container trades created an opportunity for the growth of specialized container-terminal operating companies. The companies have the resources to support substantial investments, have wide experience in container handling and logistics and have considerable expertise in technologies, particularly information technologies. The leaders globally are:

- Hutchison Port Holdings (HPH), one of the five core businesses of Hutchison Whampoa of Hong Kong, which has full ownership of seven container terminal projects, and is involved in a further 32 investment projects, which gives it international presence in 16 countries in total;
- Port of Singapore Authority (PSA), which set up an International Division in 1996 to take its expertise to other ports. It has investments in 23 port projects in 11 countries;
- APM Terminals, which is represented in 22 countries with 20 fully-owned investments and 18 further international projects;
- P&O Ports, one of the core businesses of the P&O Group, which has full ownership of eight container terminals and shared investments in a further 27 port projects, which makes up for a presence in 16 countries.

Not all successful terminal operators attempt to extend their businesses internationally. For example, ECT of Rotterdam initiated a terminal management programme for the Port of Trieste, but subsequently decided to refocus its resources on Rotterdam.

Europe has felt the direct influence of the globalization of terminal management. As HPH already owned the Port of Felixstowe, its proposed investment in ECT led to a prompt investigation by the European Commission into the effect of a dominant position in the container terminal business. P&O Ports has also been active in increasing its involvement in European ports. In February 2000, P&O Ports took over Antwerp Allied Stevedores and two large terminals from Seaport (for containers and break bulk, realizing 500,000 TEU and 1.2 million tons in general cargo in 1999). The objective of P&O Ports is to develop a pan-European network that will create synergies with other activities of the group. To this end, it also invested in the inland port of Duisburg. P&O Ports failed in an attempted acquisition of a majority shareholding or full ownership of one of the two other prominent market players (Hessenatie and Noordnatie) because of the aforementioned merger of these firms.

### 4.3 Other market players in the port

In a port environment, there are, of course, numerous other, often smaller, market players besides a port authority and terminal companies, e.g. towing services, shipping agents, hinterland modes and shippers. So far, there appears to have been little movement amongst these players, nor have they really been take-over targets. However, if the current trend towards greater control over the logistics chain persists, this may rapidly change.

Until recently, towing services in Europe enjoyed a virtual monopoly, i.e. they operated in a sheltered environment. The consequences were clear to see: a lack of investment, substandard services, and inadequate planning for the future. Towing services succeeded in achieving a degree of profitability, but often this was skimmed off by trade unions and personnel. After 1993, competition set in, with the entry of such companies as Kotug and Smit in Hamburg, and Fair Play in Rotterdam. This resulted in a sharp decline in charges.

This change has stimulated a move towards globalization. The Antwerp-based company URS, for example, is looking actively for international partnerships with a view to developing a worldwide supply of services. This implies consolidation of its operational dominance on the Scheldt and a global expansion of activities. This expansion would be achieved mostly in Latin America, not in Northern Europe, where towing services have already reached saturation point.

With regard to shipping agencies, the trend is for shipping companies to assume direct responsibility for previous agency functions. In some cases, rather than drop an agency relationship, a shipping line may enter into a joint venture with a former agency. For example, in Antwerp, the Mediterranean Shipping Company (MSC), which used to operate with a traditional agent, now has a so-called dedicated maritime agency at its disposal, which operates under the name MSC Belgium. This shipping agency is a 50/50 joint-venture between MSC and its agent. The similarity with another operation revolving around the concession for a container terminal (joint-venture between MSC and Hessenatie) is striking.

Among the hinterland transport modes, the highly competitive road transport is the most important mode. The role of railways is dependent on the provision of good dependable services. This has led to initiatives by shipping lines, terminal operators and port authorities to enter into arrangements for the operation of scheduled 'shuttle' trains.

Shippers have various influences on the patterns of vertical and horizontal reorganization. Perhaps their strongest influence is their indirect influence through the preferences that they express in the market in their purchases of transport and logistics services. Consolidation amongst the shippers is adding weight to the need for service suppliers with a global capability. Characteristics of supply-chain management are encouraging vertical integration. However, shippers may also have opportunities to express their concerns about market structures that affect competitiveness in industries. The traditional example is in relation to shipping conferences.

## 5 Conclusions

In recent years, there have been well-marked trends for vertical and horizontal integration in the international maritime logistics chain. Shipping companies in particular have been taking the initiative in this development. One consequence has been an increase in the market power of the large shipping companies over the other service suppliers, such as port authorities.

However, as expected, the other market players are responding to the changing environment also. Port authorities are adopting new strategies, and new companies are emerging in container terminal management. Mergers, regional and global expansion are changing the structure of the old stevedoring business. The large shipping companies may now find themselves negotiating with the same terminal operating company in a range of ports as well as in different port ranges. This may affect the negotiating power of the lines and increase their interest in the operation of dedicated terminals or obtaining a shared interest in a terminal. Such a trend may reduce the 'footloose' nature of shipping companies and place their relationship with terminals on a sounder and longer-term economic base.

This development would tie in perfectly with the striving of port authorities to bind their shipping customers to the port. A shipping company with a financial stake in and a long-term agreement with a dedicated terminal will be less inclined to opt for alternative ports of call. It is striking in this whole evolution how a number of port authorities are emerging as more active market players, buying themselves in to the capital of stevedores or joint-ventures between stevedores and shipping companies. The weapon they use to achieve this goal is the allocation of concessions. Could there be a better example of striving for survival?

However, these developments give rise to a number of important research questions. To what extent is the role of port authorities as a regulator and go-between with the government compromised in the eyes of (competing) market players that are not involved in the arrangement? To what extent will a further concentration of terminal management companies and the active participation of port authorities affect price determination of port services? How will the maritime markets and, more in particular, port activities evolve from now? How will the long-term commitments of shipping lines and terminal management companies to facilities affect the dynamics of competition? Answers to such questions are speculative. However, they may be aided by simulation studies to examine the influence of hinterland transport connections and the functioning of terminals on traffic flows.

### Acknowledgements

The authors express their gratitude to W. Dullaert and T. Vanelslander for their assistance in research and to G. Blauwens, H. Paelinck, F. Suykens,

M. Vermorgen and F. Witlox for the many fruitful discussions that contributed to the writing of this article. The authors remain solely responsible for the article's content.

## References and notes

1 Heaver, T., Meersman, H., Moglia, F. and Van de Voorde, E., 2000, Do mergers and alliances influence European Shipping and port competition?, *Maritime Policy & Management*, 27, 363–73.
2 Various names are given to the companies that operate container terminals. In some cases, the port authority or private port company itself is responsible for terminal management. Shipping lines manage terminals and specialist firms manage them. Some have retained the name 'stevedores', e.g. stevedoring services of America, other companies feature the word 'port', e.g. P&O ports. In this chapter, the general term 'container terminal management company' (CTMC) is used to cover all the specialist firms.
3 Heaver, T., 1995, The implications of increased competition for port policy and management, *Maritime Policy & Management*, 22, 125–33.
4 Suykens, F. and Van de Voorde, E., 1998, A quarter of a century of port management in Europe. Objectives and tools, *Maritime Policy & Management*, 25, 251–61.
5 Besanko, D., Dranove, D. and Shanley, M., 2000, *Economics of Strategy*, 2nd edn (New York: John Wiley & Sons, Inc.).

# 12 Northern range port strategy

*Jan Blomme*

## 1 Trends in the maritime and logistic sector

In the past few years the maritime and logistic sector have been shaken up under the influence of three main trends: the increase of the world trade linked to the globalisation of the economy, the unstoppable march of the container and the breakthrough of the new Information, Communication Technologies.

The increase of world trade during the 1990s is a remarkable and exceptional phenomenon, also in a longer historical context. At normal times there is a ratio of about 1.5 to 2 between the development of the GDP and the international trade (Calculated on figures of Kenwood, 1992 and OECD, 2000).[1] Therefore it is remarkable that during the last decade, international trade, compared with earlier times, increased much faster than the growth rate of the GDP. During the 1990s, the ratio between the development of the GDP and international development increased to a level of 2.5 to 3. That somewhat unexpected development is the reason why many reputable consultancy firms had to constantly re-adjust their growth forecasts with respect to the containerisation growth rates.

The main force behind this increase of growth is the globalisation of the world economy. Indeed, Global logistics is the name of the new game. By way of globalisation and the related international trade, a lot of companies aim at the realisation of important advantages of scale. They do this by concentrating their productions in the countries and the continents with comparative advantages. The components are then transported to the destination in order to be assembled and/or distributed near the important consumption areas.

The ever-growing volumes of goods that have to be transported from one continent to another, offered new perspectives to the shipping industry. It is without exaggeration to compare the super-fast march of the container in the transport business with the transition from the sailing ship to the steamship in the nineteenth century.

As a matter of fact, due to the container revolution, the cost prices of

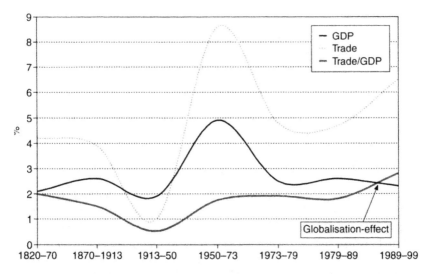

*Figure 12.1* Yearly growth (%) and relationship between GDP and Trade Industrialised World.

the intercontinental transport were profoundly reduced. Therefore, there is little doubt about it that containerisation has played an important facilitating role to make the globalisation of the word economy possible.

As a result of the globalisation of the economy, for many sectors the quality of the logistic service became much more important than before. This effect was intensified by the increase of the standard of living, which enhanced a growing individualisation of the consumer demand. The big multinationals responded to this by adding new products to their assortment, an evolution that made the logistic chain even more complex. As private or industrial activities nowadays are operating more than before on a global scale, important cost savings can be realised by diminishing the stocks (e.g. by means of JIT or SILS). As a result, logistics and transport became key elements in the competition strategy of many companies. Shippers and logistic service providers pay much more attention to integral logistic concepts comprising the whole 'production to consumer' chain. It is striking that next to cost efficiency, which is still very important, more and more attention is paid to concepts such as reliability and the certainty that delivery times can and will be respected. The breakthrough of new ICT-applications was of fundamental importance to steer the new concepts in the right direction.

However, the main question is whether this acceleration of maritime trade is structural and, if not, when it will come to an end. This is crucial, since growth rates might return to normal, this might result in huge losses for those players who invested heavily in new infrastructure.

## 2 The reaction of the main actors: concentration and globalisation

To the increasing demand for transport, the main actors reacted by growing and becoming global players themselves. The first to react were – reasonably – the shipping companies. However, this was not self evident in the strongly capital intensive shipping industry. Several strategies have been adopted in the past in order to realise the ambitious objective of globalisation. First, global alliances have been developed in order to pursue the benefits of scale economies (Lim, 1998).[2] At the same time, an increasing quality of the rendered services is envisaged by the supply of a global network, characterised by a high sailing trip frequency.

As a result of the collaboration efforts of the alliance partners, this objective could be partly attained, using a reasonably limited number of financial resources. Nevertheless, the alliances that had been developed originally proved to be characterised by an unstable structure: the collaboration between the partners was limited to the operational aspects of shipping and the different partners in the alliances often decided to follow an individual commercial strategy (Midoro, 2000).[3] As the alliances were considered not very stable and effective in order to realise more ambitious objectives, a remarkable number of mergers and acquisitions were realised after 1995. Indeed, a more fully integrated collaboration was possible by means of merging or acquiring other companies. By adopting this strategy, even some initial smaller players succeeded to become a global player (Alix *et al.*, 1999).[4]

The trend of these developments is very recent. The first large-scale alliance, the *Global Alliance*, dates back only from 1994. Concentration in the shipping industry is a phenomenon of the 1990s. In 1985, the top-20 container shipping companies controlled about one third (34 per cent) of the existing slot capacity of the world container fleet. Five years later, in 1990, their share decreased to a mere 29 per cent. Yet, in 1996, the top-20 shipping companies clearly dominated the container fleet capacity with a share of 54 per cent. The consolidation now seems to occur within the top-20 shipping companies themselves. This concentration by means of mergers and takeovers in this group is only a trend that can be observed during the last three years, starting with the mergers between Maersk and Sealand and between P&O and Nedlloyd. (see Figure 12.2).

In this 'survival of the fittest' process, a severe battle in the shipping world has started in order to become 'big, bigger, biggest, best'. For the companies that are not able to keep up with the pace of this evolution or that do not have the necessary amount of financial resources to adopt rapidly, the credo even is 'get big or get out' or maybe even more accurate: 'get big or get eaten'. It is possible that, in the end, only a handful of shipowners will hold out as players on a real global scale.

Next to the process of 'horizontal integration' that has been described

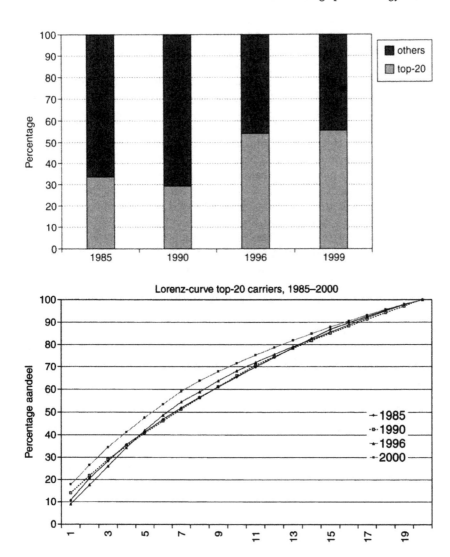

*Figure 12.2* Market share of top-20 container carriers.

above, a process of 'vertical integration' is observed since the 1990s. Here, the aim is to influence the entire logistic chain, e.g., by rendering door-to-door services, by controlling the hinterland transport and by supplying additional logistic services. The incentive behind this trend is the limited profitability of the container shipping industry. By broadening the existing and traditional set of services, the shipping companies aim to realise three objectives:

1   gaining a direct control over the quality of the product;
2   sharing commercial risks; and finally
3   realising a better relationship with their customers.

The other players in the logistic chain, the stevedores, the cargo handlers, the haulage companies and the logistic service providers, have also been confronted with these developments. With some backlog, they started to think in terms of expansions.

As far as the stevedoring industry is concerned, the share of the largest European companies increased between 1988 and 1998 from 52 per cent to 62 per cent. The increase in scale of the stevedoring companies in the Le Havre-Hamburg range has resulted mainly from the autonomous growth on the terminals they operate. Takeovers and mergers remained rather limited and were restricted to the development of satellite-companies in neighbouring ports, such as Hessenatie in Zeebrugge. An exception to this rule was ECT that decided to temporarily take over the terminal in Triëste and Antwerp-based Noordnatie which is building a jointly operated container terminal in Ventspils.

The question, however, remains how much longer the increasing concentration will be kept away from the European stevedoring picture. Eurogate, the merger between BLG from Bremen, Eurokai (Hamburg) and the Italian Contship is the first 'global' stevedoring company with an entire European network. This merger-company has already entered the top five of the world terminal operators. Eurogate also chose to broaden its scope by supplying entire logistic service packages to their customers by means of a one-stop shopping system. Here, the entire logistic chain is covered by parts of the company and also by an intermodal network for hinterland transport.

Plans exist for a merger between the two most important Antwerp stevedoring companies, Hessenatie and Noordnatie. In the near future,

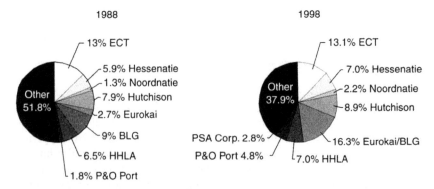

*Figure 12.3* Operators of European deep-sea container terminals 1988 and 1998.

Hessenatie will build a container terminal in Flushing and is also investigating possible expansions in the range. In this concentration, the role of the large global players – the so called 'Big Five'[5] is eminent: they have gained ground in the Mediterranean and England as a result of privatisation and of the rise of newly-developed transhipment hubs. Today, their presence in the range ports is rather limited as a result of the typical landlord structure of these ports. The allocation of new terminals is strongly monitored and regulated by the port authorities and in many cases, the global players are not very influential as yet. Consequently, the strategy of the global stevedoring companies in this region is shifting towards a takeover policy: 'if you can't join them, *buy* them' (Reyes, 2000; Davidson, 2000).[6] In this context, the most striking takeover was the partnership of Hutchison and ECT, the dominant container stevedoring company in Rotterdam. In Antwerp, P&O Ports acquired the mixed container terminal of the local stevedoring company Seaport and, as a result of the allocation procedure for new terminals, has been assigned a new terminal on the container dock that still has to be built on the Left Bank of the river Scheldt.

The process of consolidation also takes place on land with the road haulage companies and companies rendering logistic services. Companies such as Schenker/BTL and Kühne and Nagel evolved from classical forwarders to fully developed logistic service firms. In this context, the role of the large 'dinosaurs' (big and slow), i.e. the public transport companies, should be mentioned. Their importance is often neglected in the discussion: in the 1997 top-100 of the largest European transport companies, 39 entirely owned or semi-controlled public companies were listed, of which 25 were rail companies and 14 postal companies (next to 24 airline companies in which the government also often played a substantial role and further 26 firms specialised in logistic services and 11 shipping companies) (El top 100 Europeo del transporte, 1999).[7] These dinosaurs are looking for new opportunities to grow since their mere existence is threatened by the European liberalisation process. Horizontal integration of operations is not obvious because of a number of problems directly related to the national location and embeddedness of these companies. These types of companies are aiming to especially realise 'economies of scope' by means of increasing the intensity of their services along the logistic chain. Examples here are the takeover of Danzas by the Deutsche Post or TNT by the Dutch Post. Maybe the most striking example in this field is ABX, the Belgian daughter of the public railroad company, that as a result of a daring takeover campaign in the different European countries has become the third largest European logistic service provider (Sertyn, 2000 and Klotz, 2000).[8]

In this context, two final remarks have to be made. First, the exponential growth and the integration of the European transport sector might assume that within ten years time, more severe competition will exist between integrated logistic networks, controlled by a limited number of mega-players rather than between ports and local players. A second remark concerns the direction of the consolidation efforts: 'from the seas to the land', in that the shipping companies were the first to launch a vertical integration effort. Partly, this can be explained by the experience that most shipping companies gained as a result of the 'carrier haulage' approach. A more important reason is probably the high degree of specialisation that does not allow it to move in the other direction. And an even more important aspect is related to the capital intensity which is a major impediment for a different trend: the consolidation effort moves from the sea to the port and finally to the land. The large fragmentation and the limited capital intensity and turnover of many logistic firms makes these more vulnerable for consolidation. As such, these companies are the final piece of the present flood of takeovers.

### 3  Port authorities under pressure: options and reactions

In this battlefield of takeovers and mergers, the port authorities find themselves in a somewhat uncomfortable position (Heaver *et al.*, 2000).[9] Besides, it would not be correct to judge the port authorities in the Northern Range by the same standards as the other players in the maritimelogistic environment. In North-Western Europe most ports are of the landlord type, managing and developing the basic infrastructure, but exercising little activities in the commercial sphere. Their management mandate is limited to a great extent to the immediate port area. As shown in Figure 12.4, their financial strength is very restricted compared to other port players. Moreover, these port authorities in the range are working within the so called 'Hanseatic tradition', which means that almost all of them are governed by the local community – most of the time the own town government or the own city-state. The logical result is that they use a different logic than the fully privatised ports. Not simply the profit maximisation, but also the general interest for the community in the form of employment, added value creation and stability are at the centre stage of their interest. Their modest financial negotiation power often combined with institutional inertia, make them less flexible to take up new initiatives in the commercial sphere.

Generally speaking, the starting point of the port authorities' policies is the preservation and the strengthening of the competitive position of the port. The port authorities in the Northern Range are given a number of possible choices which are of vital importance to their future: how to organise the competition in the port, how to respond to the demand of the shipowners to develop dedicated facilities and finally there is the question whether or not port authorities should follow the policy of 'network building'.

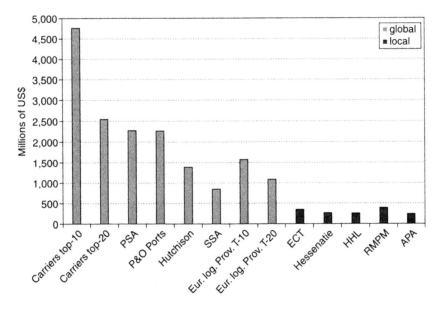

*Figure 12.4* Revenues global players as compared to local players, 1998, millions of US$.

### 3.1 The contestability-issue: the third way

As mentioned before, the prime objective of a port authority is the preservation and the reinforcement of the market position of the port they manage. Perhaps the most important responsibility of the Northern Range port authorities is the organisation of the competition aspect of stevedoring companies in the port area (Brewer, 1992).[10] Related to the contestability theory, two options are possible. On the one hand, the organisation of sufficient competition *within* the own port or on the other hand, working towards the creation of a single or dominant player who should remain alert as a result of *external pressure*. Rotterdam has opted for one dominant player in the container handling business, i.e. ECT. In the Antwerp port, until the recent past, the port authority opted for a policy designed to have several medium-sized enterprises. In the other ports of the range a combination of both approaches can be observed.

The question arises what are the advantages and disadvantages of both systems? In general, the presence of a quasi-monopolist is defended by pointing at the possibility to realise a maximum 'shareholder value', an objective that is increasingly important in case the firm studied is a local player with an evident link with the region where it is located. Major incentives for sound management are present when the geographical market, in this case the range, offers sufficient competition and alternatives to the customer. By means of external competition, exorbitant monopolistic price setting is made impossible. The port authorities, according to theory, have all interest to pursue the development of one large company. By a certain amount of market control, the new built infrastructure will be used optimally and a reasonable price level is guaranteed. Hence, sufficient means are generated for the building up of capital intensive superstructures and for investments in new technologies, making the port as a whole more attractive. In addition, by increasing its scale, the company gains the opportunity to negotiate with the other global players on equal grounds and – eventually – will become itself a global player in the market, e.g. by building up an integrated terminal network (Dijkgraaf, 1999).[11]

The most important reason why Antwerp opted for a structural internal competition in the past was the conviction that this is optimal for guaranteeing reasonable price levels for the customers. The more competitive price setting by the terminal operator allows to compensate the somewhat more expensive trip on the river Scheldt. Until recently, Antwerp was not as much a hub port in container shipping but rather a port that was called by one or more strings of the different shipping lines and their alliances. The possibility to choose among different stevedoring companies was an additional element for the shipping companies to include Antwerp in their sailing schedule.

Despite, or because of, the dominant position of ECT, the 1990s were characterised by impressive technological innovations bearing however a large degree of risk. The 'local embedding' strategy of companies was not realised. ECT was partly merged by one of the 'Big Five' global stevedoring companies. A participation of the port authority was necessary to monitor the attainment of the broad societal objectives of the company. The ambitions to build a maritime terminal network were postponed following the problems with respect to the takeover of the terminal in Trieste.

On the other hand, the 'cut throat' competition in the port of Antwerp has also provided a number of negative effects. The financial returns of the large stevedoring companies were characterised by fierce pressures. However, internal competition resulted in an unsurpassable price/quality level of the service. The Antwerp stevedores became 'lean but mean', but also largely remained local actors.

The most important lesson to be learnt, is that the concept of 'external competition' needs to be handled with caution. External competition in the range is less fierce than can be expected at first glance. The container hinterland of most of the large container ports is limited to a few hundred kilometres, often even less. For containers, the range can be considered as consisting of three parts: the Northern ports with Bremen and Hamburg, the Western ports with Antwerp and Rotterdam and in the South the port of Le Havre. In each of these regions, a task assignment between twin ports exists: Rotterdam and Antwerp on the one hand and Bremen and Hamburg on the other hand. For the Antwerp port authority the developments of the past years have provided a strong signal to change its policy on competition. It was the intention to support the creation of a large stevedoring company with sufficient 'scale and scope', but at the same time to uphold a certain degree of internal port competition. By creating a local 'oligopoly' – the third way – the port authority aims to combine the objectives of the Antwerp as well as Rotterdam port policy. This involves the creation of a limited number of (container) stevedoring companies with sufficient financial resources to grow and with a stronger negotiation position to the customers. However, at the same time, port internal competition needs to remain intact, though less fragmented than before. Essentially, the final goal is to create a situation where the container shipping companies always have an alternative available within the port area so that in case of an eventual disagreement between the shipping company and the stevedore, this should not result in a loss of cargo for the port as a whole.

### 3.2 How to deal with the shipowners and the global stevedores?

As mentioned before, obtaining dedicated facilities in the main ports is the first step for most large shipowners in their pursuit for more control on their service level. The logic behind this is that one of the main factors in controlling and improving the efficiency of the shipowner is exactly the service offered by the stevedore (Reyes, 2000).[12] Also for the global stevedores, building a terminal network in all main ports is a fundamental option.

Contrary to other maritime areas, global shipowners and global stevedores experience more difficulties to enter the Range-ports. Local players have already taken the terrain. Why should port authorities offer possibilities to newcomers if the local stevedoring companies are neither in technology nor in efficiency inferior to the global stevedores? The limited value added the global stevedores can offer, is the main reason for the thus far rather weak penetration of these companies in the Range area.

Granting dedicated facilities to shipowners, however, can generate added value for the port authority. Shipowners like MSC and Maersk/Sealand have a market share of more or less a quarter in Antwerp and in Rotterdam (Notteboom, 2000).[13] The main advantage is that through the capital investment important cargo flows are better anchored in the port where the ship owner invested in a terminal.

In Antwerp however, dedicated or semi-dedicated terminals are only granted under specific conditions. It is, for the port authority, of huge importance that the candidate has sufficient market size so that the dedicated facilities will not be underutilised. A second condition is that the handling capacity for other shipowners should not be jeopardised. A third condition is that the substantial know-how of the local stevedores is valorised through joint ventures.

There are other differences between Rotterdam and Antwerp. In Rotterdam, the first dedicated terminal owner, Maersk, will fall back on rather classical technologies. In Antwerp, the creation of a joint venture between MSC and Hessenatie is the starting point for a new wave of careful and pragmatic innovation in terminal management and superstructure. Dedicated terminals are inevitable, but the number of players eligible for these terminals is limited. Therefore, the decision of some top-20 shipowners to get involved in dedicated terminals in a limited number of ports, will just as inevitably strengthen and speed up the development of port networks.

### 3.3  To grow or to die?

A frequently asked question on port strategy is whether or not port authorities are to become commercial players and whether or not they should aim at growth outside the classical port activities.

To answer this question, the same arguments as the private sector could be used. Port networks with the active involvement of the port authority might offer a certain added value to the port user. In addition, the management risk of the port authority might be spread in a more balanced way.

But a more intense involvement of semi-public companies in setting up port networks is nevertheless not straightforward, neither on the maritime side, nor on the landside. To start with, it is rather contradictory to have initiatives in this sense, at a time where about everyone agrees that the public sector should get out of commercial activities. The involvement of port authorities are, in the end, financed with taxpayers' money, and investments in port-related activities are, needless to say, not completely without risk. And, last but not least, there is the problem of possible conflicts of interest. Port authorities in main ports are expected to respect the principle of equality. The risk of conflicts of interest with other port users, inside or outside the port, is very real. At present, new public owned players (partly) financed with public-money enter the market. They are regarded with distrust by private companies. And who will prove them wrong?

This argumentation however needs to be put in the proper perspective. Most ports in the Range developed a holding structure in order to differentiate between their port management functions and their commercial activities. Moreover they could play an important role as a neutral player in developing and optimising or supporting services such as port-related ICT.

For other strategic goals, such as hinterland accessibility, the port authority can be an active partner for the private sector. Most initiatives so far in North-Western Europe failed however because of the local autonomy. But this is most likely a question of time.

Finally, let us not forget that most port authorities will be, for the next decade, occupied by their own mega-projects. To 'grow or to die' is not necessarily the same as establishing port networks. In most ports there is an important local potential for growth. The tidal container dock and the Verrebroek Dock in Antwerp, the second Maasvlakte in Rotterdam, the Africa port and the Ceres terminal in Amsterdam, the Altenwerder terminal in Hamburg, the plans of Hamburg and Bremen for a container terminal in Wilhelmshaven and the 2020 project in Le Havre will require a lot of energy and means to be invested by the respective port authorities. Accommodating the expanding container market is actually for many port authorities the first priority. And wanting to grow too hard too fast is not entirely without risk.

## 4 Conclusions

The concentration trend in the maritime and logistic industry has created some distortions in terms of market power. Shippers, import and export companies as well as consumers benefited from these developments. However, a number of smaller players among the shipowners and the logistic providers will not hold their own or will have to follow a niche-strategy to survive. In Antwerp, an important goal is to give way to a limited number of strong local stevedores so as to maintain the principle of internal competition. 'Dedicated' terminals are not rejected since they anchor important commodity flows in the port, but only granted under specific conditions. Collaboration and networking on the level of the port authorities still is, for a variety of reasons, in its infancy.

## References and notes

1 Kenwood, A.G. and Lougheed, A.L., 1992, The growth of the international economy, 1820–1960. An introductory text, London/Sydney, OECD, Perspectives Economiques, 1985–2000.
2 Lim, S.M., 1998, Economies of scale in container shipping, *Maritime Policy Management*, 25, 4, 361–73.
3 Midoro, R. and Pitto, A., 2000, A critical evaluation of strategic alliances in liner shipping, *Maritime Policy & Management*, 27, 1, 31–40.
4 Alix, Y., Slack, B. and Comtois, C., 1999, Alliance or acquisition? Strategies for growth in the container shipping industry, the case of CP ships, *Journal of Transport Geography*, 7, 203–8.
5 Hutchison, PSA, P&O Ports, SSA and ICTSI.
6 Reyes, B., 2000, Pros and cons of port ownership', N. Davidson (ed.) *Lloyd's List*, 23/08/000, Gap widens between key international players, *Lloyd's List*, 23/08/00.
7 El top 100 Europeo del transporte, 1999, *Logistica, Transporte, Paqueteria y Almacenaje*, 15–16.

8 Sertyn, P. and Boval, Ch., 2000, We zijn een transportbedrijf, geen spoorweg-maatschappij (interview met E. Schouppe), *De Standaard*, 21/06/00, 37. Klotz, H., SNCB streifen das Schienenkorsett ab, *DVZ*, 27/07/00.

9 Heaver, T., Meersman, H., Moglia F. and Van de Voorde, E., 2000, Do mergers and alliances influence European shipping and port competition? *Maritime Policy & Management*, 27, 4, 363–67.

10 Brewer, P.R., 1996, Contestability in UK rail freight markets. The economies of open access, *Transport Policy*, 3, 3, 91–8.

11 Dijkgraaf, E., Haffner, R.C.G., van der Schans, P.T. and Varkevisser, M., 1999, Handen af van ECT?' *ESB*, 10–13. Dit artikel is gebaseerd op *De rol van de overheid in de containeroverslagmarkt*, Onderzoekscentrum Financieel Economisch Beleid, Research Memorandum 9805, Rotterdam (1998).

12 Reyes, B., 2000, Op. Cit.

13 Notteboom, Th., 2000, *De invloed van ruimtelijke en logisiteke ontwikkelingen in het voorland-achterlandcontinuüm op de positie en functie van zeehavens*, doctoraatsverhandeling, RUCA.

# 13 The peripheral port challenge in container port systems

*Theo E. Notteboom*

## 1 Introduction

The container and intermodality concept resulted in a metamorphosis, characterised by technological innovations and changes in the organisation and structure of the distribution system. New patterns of freight movements have emerged as a consequence of the changes in port hierarchy and the extension of ports' hinterlands. Since the beginning of the 1990s, European containerisation has entered a new phase as shipping lines exert a huge demand pull on seaport systems. The requirements imposed by shipping lines on deep-sea container terminals have changed considerably, not only in terms of productivity, handling costs and service, but also in terms of location-bound elements such as the maritime access profile and the proximity of the cargo-generating hinterlands.

This contribution explores some of the questions related to the development of container port systems and more in particular the rise of new terminals and former non-hub terminals at the expense of the existing large load centres. The concept of the 'peripheral port challenge' will be used to analyse deconcentration patterns in port systems. The central hypothesis put forward is that the peripheral port challenge is an inherent ingredient of contemporary port system development. This hypothesis will be tested on the situation in the Rhine-Scheldt delta port cluster and the West Mediterranean container port range. Furthermore, this paper will elaborate on the rationale behind deconcentration tendencies within container port systems.

## 2 A conceptual approach with respect to the peripheral port challenge

Most theoretical models on seaport system development suggest that large ports, which invested early in container infrastructure, attract more and more container traffic. The resulting port concentration can cause degradation of minor ports in the network.[1] Only few spatial models refer to a deconcentration tendency caused by the growth of former non-hub ports and the emergence of new ports. A distinction should be made between deconcentration within a port and deconcentration within a port system (Figure 13.1). The first type basically refers to the infrastructural extension of port areas away from the historical core to less urban sites. This kind of downstream development of terminal infrastructures is highlighted in the 'Anyport' model of Bird,[2] the port city model of Hoyle[3] and the model of Barke.[4] Scale expansion in shipping and the use of new transhipment technologies are some of the factors that have resulted in the abandoning of older port sites.

Deconcentration within a port system occurs when some of the cargo is shifted from large ports to smaller or new ports, or when the large load centres only absorb a small portion of the container growth in the port system. Hayuth calls this phenomenon the *peripheral port challenge*.[5] He argues that as the port system develops, diseconomies of scale in some large load centres emerge in the form of a lack of space for expansion and port congestion. This encourages smaller ports or even new ports to attract cargo. The peripheral port challenge concept thus implies that those ports which existed before the container revolution and invested early in the new technology are gradually losing market share to new or upgraded ports which try to gain hub status. The next section provides an insight into the changing hierarchy in container port systems and the emergence of a new generation of hub terminals.

## 3 The changing hierarchy within port systems

### 3.1 Hub-and-spoke systems in liner shipping

Hub-and-spoke networks have become a desirable form of service configuration in liner shipping. They offer great advantages for carriers in terms

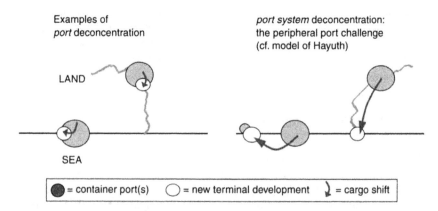

*Figure 13.1* Deconcentration patterns in port systems.

of the consolidation of cargo flows and in terms of potential market coverage.[6] Furthermore, this type of network allows considerable economies of scale of equipment. In container shipping this is materialised through the deployment of post-panamax ships.

Round-the-world services (RTW) and pendulum services are widespread forms of hub-and-spoke systems. The emphasis of such services lies on a few selected hub ports of call, linked to all the other destinations by feeder services and/or inland shuttle services. In the 1980s, some container shipping lines started to abandon traditional routing systems in favour of RTW services.[7] In the early 1990s, some carriers introduced pendulum services. The system relies on hub ports that act as turntables between liner services of two different trades. This kind of liner service design is well suited for high-volume international trade routes such as the trade Europe-Far East-US West Coast.

### 3.2 A new generation of load centres

When designing hub-and-spoke networks carriers have to decide about where to place the hubs and how to route the traffic that is to flow between the origins and destinations over the resulting network.[8] Change is overwhelming the location, size and productivity of ports around the globe, as the container industry prepares for the full impact of the deployment of a new generation of container vessels. As such, the last decade has seen the emergence of a new breed of hub terminals along the east-west main shipping lanes. Traditional container ports around the world have a shared frustration as they observe the development of new container hubs at unlikely places far away from the immediate hinterland that historically guided port selection. Sites are being selected to serve continents, not regions; for transhipping at the crossing points of trade lanes; and for potential productivity and cost control. Using the terminology of Hayuth and Fleming,[9] the new hubs rely heavily – sometimes completely – on traffic flows that are distantly generated by the interaction of widely separated places and stimulated by the port's en route location or intermediacy. Such intermediacy-based container flows strongly support the 'cargo follows ship' argument.

### 3.3 Port competition and the rise of intermediacy-based traffic

The dynamics in liner service network design increased the relative importance of intermediacy-based flows, also in the traditional large load centres. The growing importance of the 'cargo follows ship' argument is partly the result of the scale increase in vessel size and the observed concentration in liner shipping through alliances, mergers and acquisitions. The increase in intermediacy-based container traffic has broadened port competition:

- The geographical scale for accommodating intermediacy-based flows covers an entire port range (e.g. the Hamburg-Le Havre range) or sometimes even an entire continent (e.g. the European container port system). As the latter type of traffic gains in importance, load centres located far way from each other are now competing. Notteboom[10] demonstrated that competition among load centres in the Hamburg-Le Havre range and the Mediterranean range has intensified considerably in the last couple of years.
- The pressure of medium-sized load centres and new hub terminals on the position of traditional load centres is increasing.

### 3.4 Diversion distance as a determining factor in liner service network design

Many port operators and shipping lines have understood that the specific requirements related to the current trends in the design of deep-sea services do not necessarily make the existing large load centres the best locations for setting up hub operations. The result is that alternative terminal sites, either in small ports and or new port areas in the periphery of existing load centres or along coastlines with favourable draft conditions, draw a lot of attention in port and shipping circles.

The concept of *diversion distance* is often used in this context. The diversion distance can be defined as the one-way distance a mother vessel has to sail to get from the main maritime lane to the desired hub port of call. The scale increases in vessel size and with it the increased focus on fast turnaround times put more pressure on the reduction of the diversion distance. The minimisation of the diversion distance is particularly important when intermediacy-based transhipment flows (relay or feeder) are involved. The shipping lines' focus on small diversion distances and high available drafts makes that new terminals are developed along deep coastal waters or downstream of a main river. A peripheral port challenge emerges in case these new large terminals are built on new port sites, in former non-containerised ports or in smaller container ports.

### 3.5 Upstream versus downstream ports

As many traditional load centres at the same time act as sea-sea interfaces and sea-land interfaces an upstream location is a blessing for some types of container flows, but a disadvantage for others. A downstream location in many cases offers a better maritime accessibility in terms of access time to the port in combination with the diversion distance and available draft.[11] However, a downstream location often faces some drawbacks in the area of inland transportation, certainly if you consider that inland transport costs per tonne-km are much higher than in maritime transport.[12] Although the discussion on downstream versus upstream load centres can not be generalised, there still exists a competitive potential for upstream urban ports to function as a direct port of call in RTW and pendulum services.

The carrier's choice between a direct call at an upstream port with the mother vessel or an indirect call via a feeder vessel is not only determined by the diversion distance, but also by the volume of containers involved, the related costs, port productivity and the strength of the individual carrier in the markets served.[13] A direct call at an upstream port often is justified – even if the mother vessel carries a lot of transhipment traffic – when the carrier can combine transhipment activities with a strong cargo-generating power of the port's regional hinterland. In that case the port has more chance of generating the critical mass necessary to justify a direct call. Moreover, the position of an upstream port in linehaul service networks has more chance of being sustainable if the port succeeds in outper-forming downstream ports in terms of terminal productivity, prices and integrated value-added services, all this in order to compensate for the extra sailing time. Draft limitations remain the worst threat to the position of upstream ports in some of the main liner service networks. Many upstream ports have responded to the realities in the liner market by engaging in extensive dredging programmes to guarantee easy access to post-panamax vessels with a design draft of 13.5 to 14 metres or a commercial draft of some 12 to 12.5 metres.

### 3.6 The response of market players to the renewed port hierarchy

Despite low margins and fierce competition, the success of some hub terminals and the sustained growth of containerisation have attracted the attention of private and public investors. Large stevedoring companies, some of them operational on a world scale (e.g. P&O Ports, Hutchison Port Holding, PSA Corporation and Eurogate) others on a more regional scale (e.g. Hessenatie), have responded to the market pull by designing and constructing new hub terminals. Alternatively, an increasing number of new hub terminals is controlled by large container lines either by direct ownership, long lease or in the form of a joint venture with local stevedoring companies. This falls in line with their objective of controlling costs and co-ordinating operations throughout the entire transport chain.

The next section provides an in-depth analysis of the changing port hierarchy in two port systems in Europe: the West Mediterranean port range and the Rhine-Scheldt delta cluster.

# 4 Analysis of the peripheral port challenge in selected European port systems

## *4.1 The West Mediterranean port range*

The West Mediterranean, once considered as a niche market, is now being promoted as the back door of Europe. In recent years there has been a tremendous growth in the region's container throughput (see Table 13.1). Only recently, major container shipping lines have started to include Mediterranean ports directly in their liner services. New Mediterranean ports have emerged to accommodate the pendulum and point-to-point services with the best technology and location, whilst at the same time medium-sized ports are reinforcing their position.

Many vessels have to pass through the Mediterranean as part of their routing in order to transit the Suez Canal. Under these circumstances the opportunities to interline multiply. Consequently, many carriers are using Med ports to shift boxes between linehaul services in order to serve more markets with fewer vessels. Transhipment hubs such as Gioia Tauro, Marsaxlokk and Algeciras have a small local market (less than 10 per cent of its traffic) and are thus almost wholly dependent on transhipment cargo (relay and feeder). The success of these ports is partly the result of the fact that a call involves a minimal diversion for mainline vessels transiting the Mediterranean. Location factors, i.e. the proximity to the main shipping lane, seem to be the primary reasons for the emergence of new Med ports.

*Table 13.1* Container throughput in the West Mediterranean port range

| | Diversion distance in nautical miles | Container throughput in 1,000 TEU | | | | |
| --- | --- | --- | --- | --- | --- | --- |
| | | 1975 | 1988 | 1995 | 1999 | 2002 |
| Gioia Tauro | 67 | 0 | 0 | 16 | 2,253 | 2,955 |
| Algeciras | 1 | 50 | 412 | 1,155 | 1,833 | 2,229 |
| Valencia | 142 | 51 | 343 | 672 | 1,153 | 1,817 |
| Genoa | 355 | 162 | 327 | 615 | 1,234 | 1,531 |
| Barcelona | 211 | 78 | 410 | 689 | 1,235 | 1,421 |
| Marsaxlokk | 6 | 0 | 8 | 515 | 1,100 | 1,230 |
| La Spezia | 339 | 17 | 282 | 965 | 843 | 1,000 |
| Marseille | 292 | 94 | 390 | 498 | 664 | 813 |
| Leghorn | 301 | 66 | 479 | 424 | 479 | 545 |
| Taranto | 95 | 0 | 0 | 0 | 0 | 472 |
| Naples | 156 | 20 | 128 | 207 | 334 | 446 |
| Venice | 600 | 6 | 91 | 123 | 200 | 263 |
| Trieste | 639 | 34 | 114 | 152 | 189 | 185 |
| Ravenna | 472 | 50 | 166 | 193 | 173 | 160 |
| Savona | 356 | 3 | 18 | 47 | 25 | 55 |
| Tarragona | 189 | 0 | 15 | 36 | 40 | 53 |
| Sète | 300 | 15 | 22 | 10 | 8 | 20 |
| Total | – | 646 | 3,208 | 6,317 | 11,763 | 15,195 |

Figure 13.2 provides the results of a net shift analysis applied to three groups of Med ports. The grouping is based on the diversion distance to the main shipping lane. The net shift reflects the total number of containers a port has actually lost to or won from competing ports in the West Mediterranean range with the 'expected' container traffic as a reference. The 'expected' throughput is based on the assumption that the considered port would simply maintain its market share and as a consequence would evolve in the same way as the port range as a whole (same growth rate as the range).

Figure 13.2 demonstrates that the hub battle partly shifted activities from remote ports, in terms of the diversion distance, to nearby ports. Since 1994 remote load centres (>250 nautical miles) lost a potential growth of some 200,000 to 350,000 TEU per year mainly to the hubs Algeciras, Gioia Tauro, Marsaxlokk and newcomer Taranto. The net shift in TEU is equivalent to a considerable missed annual growth of the remote ports of 13.2 per cent in the period 1994–97 and 6.5 per cent in the last period of observation. Will this shift continue in the future? Most likely it will, but there are limits to this trend. A lot of new terminal initiatives are being developed along the main shipping lane, whereas at the same time many other hubs along the same lane are in the process of upgrading their facilities to handle more traffic. This observation supports the idea of the peripheral port challenge for transhipment traffic. For instance, Italy has three major transhipment hubs with a low diversion distance: Gioia Tauro, Cagliari and Taranto.

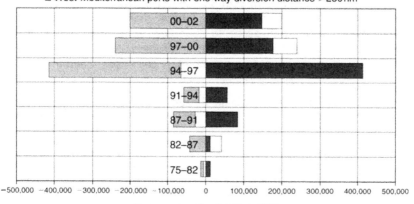

*Figure 13.2* Average annual net shifts in TEU between categories of Mediterranean container ports based on the one-way diversion distance (period 1975–2002).

The question is whether there is enough room for more hubs in the region. There are fewer lines left making direct calls to convince of the benefits of transhipment. Each of the hubs will, therefore, have to compete strongly with each other. Moreover, the increase in the number of Mediterranean hubs puts a downward pressure on the terminal occupancy. The industry expects that the average terminal occupancy will decrease from 85 per cent in 2001 to some 65 per cent in 2004 as a consequence of the new terminal initiatives.

It can be concluded that the West Mediterranean port system is a prime example of a peripheral port challenge triggered by the need to decrease diversion distances for the time-sensitive post-panamax vessels. It is expected that the observed container shifts away from remote ports will therefore eventually come to a halt. Most likely, this will coincide with a freezing or possibly a rationalisation in the number of transhipment hubs along the main shipping lane. Transhipment hubs will have to respond to footloose carriers and associated high traffic volatility via the development of value-added activities and, if geographically feasible, an increased focus on the sea-land interface.

### 4.2 The Rhine-Scheldt delta port cluster

The Rhine-Scheldt seaport system consists of a number of small and medium-sized ports and two main ports in the full meaning of the word, i.e. Rotterdam and Antwerp. Rotterdam, Antwerp and to a lesser extent Zeebrugge are involved in the deep-sea container business. In 2002, these ports handled over 12 million TEU as indicated in Table 13.2. Rotterdam is a downstream port at the estuary of the Rhine, Zeebrugge is a coastal port and Antwerp is a river port situated upstream of the river Scheldt. Rotterdam can accommodate the largest container vessels at the Maasvlakte. The maritime access to Zeebrugge allows ships with a draft of 55'. In Antwerp, draft conditions have substantially improved after the completion of a dredging programme. Only the largest container vessels

Table 13.2 Container throughput in the Rhine-Scheldt delta container port system

| | Total container throughput in 1,000 TEU | | | | | | Average annual growth | |
|---|---|---|---|---|---|---|---|---|
| | 1975 | 1988 | 1992 | 1995 | 1999 | 2002 | 1975–2002 | 1992–2002 |
| Rotterdam | 1,079 | 3,289 | 4,123 | 4,787 | 6,345 | 6,515 | 6.9% | 4.7% |
| Antwerp | 297 | 1,470 | 1,836 | 2,329 | 3,614 | 4,777 | 10.8% | 10.0% |
| Zeebrugge | 184 | 246 | 526 | 529 | 850 | 959 | 6.3% | 6.2% |
| Amsterdam | 32.1 | 66.4 | 71.8 | 91.1 | 46.2 | 45.0 | 1.3% | −3.2% |
| Ghent | 10.5 | 8.8 | 9.4 | 5.8 | 11.0 | 21.3 | 2.7% | 8.5% |
| Flushing | 27.9 | 24.7 | 6.0 | 6.0 | 2.2 | 9.3 | −1.9% | 4.5% |
| Total | 1,631 | 5,104 | 6,572 | 7,747 | 10,869 | 12.327 | 7.8% | 6.5% |

still need to take into account some tidal window restrictions. In contrast to the Mediterranean hubs, the Rhine-Scheldt delta port cluster can rely on a strong cargo-generating direct hinterland. A large share of the inland flows has its origin or final destination in one of the neighbouring countries. About 74 per cent of Rotterdam's transit is generated in Germany, Belgium and the UK. For Antwerp and Zeebrugge the overall share of the neighbouring countries (France, Germany, the Netherlands and the UK) exceeds even 90 per cent.

As indicated earlier the hypothesis of the existence of the peripheral port challenge is confirmed if the large load centres (i.e. Antwerp and Rotterdam) are losing market share to smaller or new ports. The annual net shifts in Figure 13.3 do not confirm the existence of a peripheral port challenge. The net shifts for the combination Antwerp-Rotterdam are very small in comparison to its container throughput. The negative shifts in the periods 1991–94 and 1997–2000 are marginal and do not provide a firm proof of consecutive substantial losses for the large load centres.

Do coastal ports gain market share at the expense of upstream ports? Based on Figure 13.4 it can be concluded that up to now upstream ports, i.e. basically the Antwerp terminals, have gradually gained market share at the expense of the other load centres. The net shifts in the period 2000–02 are quite substantial: a positive shift of 184,000 TEU on a total throughput at upstream ports Antwerp and Ghent of about 4.1 million TEU in 2000. The diversion distance is measured from the point where the vessel has to leave the main maritime lane in the North Sea. For the ports in the Scheldt

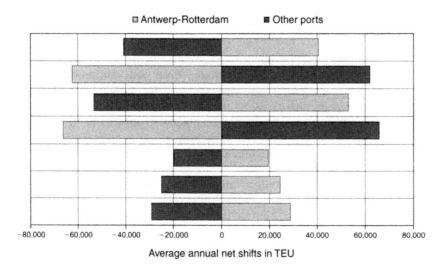

*Figure 13.3* Average annual net shifts in TEU between the large load centres Antwerp and Rotterdam and other ports in the Rhine-Scheldt delta (period 1975–2002).

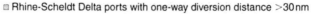
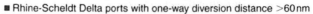
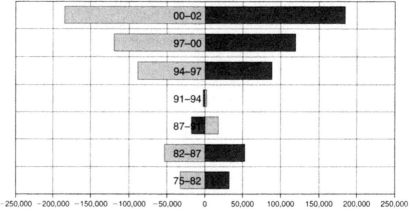

*Figure 13.4* Average annual net shifts in TEU between the upstream and downstream ports in the Rhine-Scheldt delta (period 1975–2002).

basin (i.e. Flushing, Ghent, Zeebrugge and Antwerp) the A1-buoy in the North-Sea marks this spot.

The observation that 'remote' ports in terms of diversion distance gain market share is opposite to the findings for the Mediterranean range. This does not come as a complete surprise:

*   As indicated earlier, the Rhine-Scheldt seaport system has a strong focus on centrality-based container flows and intermediacy-based flows with an inland leg, whereas the distribution system in the Mediterranean range is more organised on a hub-feeder basis.
*   Antwerp's forwarding sector (some 300 companies) adds to its strong cargo-generating power. The availability of return cargo serves as one of the key reasons why container carriers call at Antwerp, despite the upstream location of the port.
*   A survey on ship operators' image of and satisfaction with European container ports, revealed that Antwerp is considered as one of the best performing load centres in northern Europe.[14] The high level of satisfaction is the result of the port's elevated productivity, its integrated service package, the very competitive rates and the good hinterland connections. Studies of Marconsult confirm these conclusions.

The above findings do not entirely exclude a pressure for the minimisation of the diversion distance. The large load centres in the area have responded in an adequate way to this challenge by the downstream development of new terminals. This strategic decision was also triggered by the need for large terminal surfaces and the limited availability of space in the existing port areas.

In the 1980s, the sustained growth of container throughput in Rotterdam led to the construction of massive container facilities on the Maasvlakte, an area that was reclaimed on the sea. The Maasvlakte terminals handled 3.5 million TEU in 1999 or approximately 55 per cent of Rotterdam's container throughput. In 1988 this share was only 18 per cent. The Maasvlakte clearly is an example of a deconcentration tendency within a port area.

Antwerp has witnessed the same kind of development in the 1990s. The relative decline in importance of the dock system behind the locks (as depicted in Figure 13.5) is explained by the strategic decision of the Antwerp port community and the Flemish government to build container capacity along the river Scheldt in front of the locks, thereby allowing considerable savings in the port turnaround time of container vessels. The first Scheldt terminal (Europaterminal) started operations in 1990. The second Scheldt terminal (Noordzeeterminal) followed in 1997.

The port authorities of Antwerp and Rotterdam will further develop downstream port areas in the future. The port of Rotterdam has developed ambitious infrastructure plans to build a second Maasvlakte in the sea of which a part would be dedicated to the container business. To meet the container demand in the future, Antwerp is building a tidal

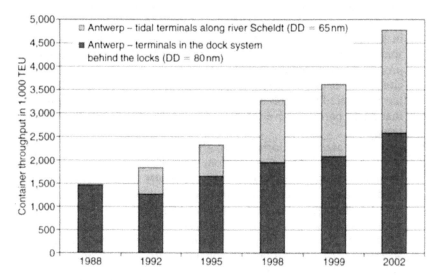

*Figure 13.5* Spatial distribution of the container throughput in the Antwerp port area (DD = diversion distance).

container dock on the Left Bank of the River Scheldt to be completed in three successive phases. The new dock will then consist of 4.85 kilometres of quay wall and cover an area of 230 hectares. The first and second phase should be ready by mid-2005. The tidal dock should reach an annual capacity of six million TEU after completion.

In the medium term, Antwerp and Rotterdam will undoubtedly face competition from new load centres in the immediate vicinity:

- The relatively new coastal port of Zeebrugge is determined to attract more intercontinental container traffic, by manifesting itself as an alternative for the large load centres.
- The Dutch seaport Amsterdam opened the Ceres Paragon terminal in 2002 with a capacity of some 950,000 TEU. A state-of-the-art handling system based on a 'slip' concept should allow Amsterdam to develop a transit role. Up to now the terminal remained empty.
- The port of Flushing is planning to become the home of the Wester-scheldt Container Terminal with an initial capacity of 1.75 to 2.5 million TEU and a quay length of 2.6 kilometres. The Antwerp stevedoring company Hesse Noordnatie would operate the terminal and would make the necessary investments in the terminal superstructure (cranes, etc.).

The upcoming peripheral port challenge in the Rhine-Scheldt delta is mainly the result of:

- the strong and sustained growth of containerisation in the region;
- the danger of terminal capacity constraints in Antwerp and Rotterdam in the medium term. Both load centres have come up early with a solution (respectively Containerdock West and Maasvlakte II) but the process towards a formal approval of these projects in government circles has been extremely slow. Other ambitious ports in the area have benefited from this situation by presenting new terminal projects that might partly resolve the region's capacity constraints in the short term;
- Antwerp-based stevedoring company Hesse Noordnatie, now 100 per cent owned by PSA Corporation, has extended its operations to coastal ports in order to offer the customers a more differentiated product range. Carriers have the choice between a call at a coastal port (Zeebrugge or in the future also Flushing) with onward inland services to Antwerp, or a direct call in Antwerp;
- the coastal ports are also partly an answer to the restricted maritime accessibility of Antwerp and the ongoing debate on the further deepening of the river Scheldt.

# 5 Conclusion

In general, the economies of scale linked to containerisation are believed to enhance the concentration of large volumes of containers in few large load centre ports. It is generally true that containerisation leads to port concentration, but it is not a confirmed fact that the container volumes will be only directed towards the existing large load centres that have invested early in container technology. Hayuth suggests that the concentration tendency in port systems will eventually reach a limit or might even develop into deconcentration because of the peripheral port challenge.

New forms of liner service network design have changed the hierarchy in port systems. A striking evolution has been the emergence of a new generation of hub ports along the east-west trunk route. In the Mediterranean this has led to the rearrangement of the distribution system around a few transhipment hubs characterised by low diversion distances. The focus of these hubs lies on intermediacy-based sea-sea container flows. It was demonstrated that the West Mediterranean port system is a prime example of a peripheral port challenge triggered by the need to decrease diversion distances for the time-sensitive post-panamax vessels. However, the observed container shifts away from remote ports will eventually come to a halt. Transhipment hubs will have to respond to footloose carriers and associated high traffic volatility via the development of value-added activities and, if geographically feasible, an increased focus on the sea-land interface.

Up to now the large load centres in the Rhine-Scheldt delta area have not been confronted with a peripheral port challenge. But also in this case the new requirements of liner shipping and the sustained growth in containerisation have attracted new or smaller ports. In five years time we will be able to judge whether the market has reacted positively to the new terminal initiatives and, as such, has encouraged a peripheral port challenge in the port cluster.

In any case, the problem associated with the peripheral port challenge phenomenon is that non-hub ports lack the scale and cluster effects of large load centres. It normally takes a while to get the necessary critical mass that allow new load centres to set up frequent and competitive inland shuttles and to offer regular sailings to a wide range of overseas destinations. In an attempt to attract the necessary critical container volume, new terminals may have to invest in facilities that may not be immediately financially sustainable. The resulting danger of structural excess capacity in terminal facilities can put a downward pressure on terminal rates and gives shipping lines and alliances more opportunities to use their bargaining power to play off one port against another.

Another danger is that new terminals might try to overcome the initial lack of scale by adjusting the price setting for inland services or feeder services, in addition to existing port equalisation systems of carriers. Shipping lines often use port equalisation systems to ensure that shippers are compensated for possible cost disadvantages in hinterland transportation towards new terminals with a favourable nautical location.[15]

Co-operation and co-ordination among the various ports will prove to be very important in view of avoiding destructive port competition. The existing large load centres can adopt a leadership role in this process. Load centres are facing more competition at different levels, but at the same time it is clear the sentence 'port needs port' is more relevant than ever. Sea ports are becoming increasingly interrelated with other ports and inland centres. It is a question of finding the right balance between competition and co-operation in order to achieve a sustainable competitive advantage for both the individual load centres in a port system and the system as a whole.

A special task may fall on smaller terminals. On the one hand there are not the volumes that justify further investments on a par with the major terminals, while on the other hand they can hardly sit back without continuing to develop and modernise their facilities. Smaller terminals should focus on technical advancements that help increase productivity and improve services, but are also suitable for their business sector, their customers' volumes and vessel sizes. Smaller terminals should also fulfil their equally important role as local terminals rather than trying to compete with, or copy the role of major hubs. Incentives to feeder operators and intermodal operators must form a part of their strategy in order to better connect the port facilities to the hubs' extensive water and land networks.

### References and notes

1 Taaffe, E.J., Morrill, R.L. and Gould, P.R., 1963, Transport expansion in underdeveloped countries: a comparative analysis, *Geographical Review*, 53, 503–29.
2 Bird, J., 1971, *Seaports and Seaport Terminals*, Hutchinson University Library.
3 Hoyle, B.S., 1998, The redevelopment of derelict port areas, *The Dock & Harbour Authority*, June 1998, 46–9.
4 Barke, M., 1986, *Transport And Trade* (Edinburgh: Oliver & Boyd).
5 Hayuth, Y., 1981, Containerisation and the load centre concept, *Economic Geography*, 57, 160–76.
6 Slack, B., 1999, Satellite terminals: a local solution to hub congestion?, *Journal of Transport Geography*, 7, 241–6.
7 Lim, S.M., 1996, Round-the-world service: the rise of Evergreen and the fall of US lines, *Maritime Policy & Management*, 23, 119–44.
8 O'Kelly, M., 1998, A geographer's analysis of hub-and-spoke networks, *Journal of Transport Geography*, 6, 171–86.
9 Hayuth, Y. and Fleming, D.F., 1994, Concepts of strategic commercial location: the case of container ports, *Maritime Policy & Management*, 21, 187–93.
10 Notteboom, T., 1997, Concentration and load centre development in the European container port system, *Journal of Transport Geography*, 5, 99–115.
11 Baird, A.J., 1996, Containerisation and the decline of the upstream urban port in Europe, *Maritime Policy & Management*, 23, 145–56.

12 Notteboom, T., Coeck, C., Verbeke, A. and Winkelmans, W., 1997, Containerisation and the competitive potential of upstream urban ports in Europe, *Maritime Policy & Management*, 24, 285–9.
13 Zohil, J. and Prijon, M., 1999, The Med rule: the interdependence of container throughput and transhipment volumes in the Mediterranean ports, *Maritime Policy & Management*, 26, 175–93.
14 Thomas, B.J., 1997, The brand equity of European seaports: a survey on ship operators' image of and satisfaction with European seaport authorities and container terminal operators, *Port Finance '97 Conference*, London.
15 Gilman, S., 1997, Multimodal rate making and the structure of container networks, *Essays in Memory of Professor B.N. Metaxas*, University Of Piraeus, 327–35.

# Part IV

# Logistics and ICT

*Edited by Pietro Evangelista*

# 14 Innovating ocean transport through logistics and ICT

*Pietro Evangelista*

## 1 Introduction

As markets become more turbulent, product life cycles shorten, product varieties increase and customer demands escalate, companies have to find new ways to sustain their competitiveness. A successful strategy to develop a competitive edge in most industries is to stay close to customers and continuously adapt products and services in response to their specific needs. In the transport and logistics service sectors, changes in customer demand represent one of the main drivers of development.

Of relevance within this context is the growing trend towards the widespread adoption of the supply chain management (SCM) concept by shippers (i.e. manufacturers and retailers). In the last ten years, SCM has become an increasingly important success factor as companies face the challenges of product proliferation on a global scale, rapidly changing technologies, the need for improved integration of functions across enterprises, and battle for market share while improving shareholder value. SCM has the potential to create value beyond simply lowering transportation costs or improving labour or distribution productivity enhancing companies' capital efficiency (i.e. through reducing inventory and facilities investment) and integrating supply chain processes.

Since the early 1990s, the literature on the SCM approach has featured a plethora of academic works. According to Christopher[1] 'leading edge companies have realised that the real competition is not company against company, but rather supply chain against supply chain,' SCM has been defined as 'an integrative philosophy to manage the total flow of a channel from earliest supplier of raw material to the ultimate customer, and beyond, including the disposal process'.[2] It is based on the following three ideas: effective purchasing and distribution; a focus on long-term relationships between trading partners; and the operational integration of trading organisations. Companies are increasingly adopting a supply chain approach as a new way to integrate their own operations and to extend this integration to their supply chain partners. Co-ordination among the various actors in the supply chain is thus an important prerequisite in order to achieve competitive advantage. Such co-ordination requires a high degree of organisational integration between a manufacturer and its suppliers of goods and services.[3]

Adoption of the SCM approach by shippers has many implications. First, restructuring logistical systems and integration of supply chains require the re-engineering of physical and material flows with considerable consequences for logistics and transportation operations management.[4] Second, for shippers, the delivery system has become integral to the value of the product supplied to the point that transportation and logistics receive the same evaluation as the product itself.[5] Third, wide adoption of the supply chain view has resulted in changes in relationships among transportation providers and other participants in the supply chain. New alliances, partnerships, mergers and acquisitions have become common. Also, new types of third party logistics providers (3PLs) have become common.[6] 3PLs have developed to provide the logistics resources and skills needed by manufacturers and retailers as they strive to achieve supply chain agility while reducing overall costs. The growing outsourcing of logistics activities is also consistent with the trend to reduce the number of suppliers and establish a closer, long-term relationship with them for providing 'tailor-made' logistics and transportation services.

3PLs are innovatively transforming the scope and characteristics of their services to improve customer service levels.[7] As a result, 3PLs play a more important role than in the past insofar as they are entrusted with the task of integrating and accelerating physical and information flows along multiple levels of the supply chain.[8] This has given 3PLs a new potential role in customising supply chains as they assume responsibility for a growing number of activities beyond transportation and warehousing. For example, the practice of postponement of product finishing to downstream stages of supply chains means that 3PLs have the opportunity to offer services such as final assembly and customisation of products. Offering these services gives 3PLs the opportunity to penetrate segments of supply chains with higher added-value services compared to traditional transportation and warehousing services. The supplementary customised services can give a differentiation edge, while raising added value in services can improve margins, as well as deepen the relationships with customers. The transition from the traditional 'arms length' approach to the supply of integrated logistics services packages on 'one-stop shopping' base and has further fuelled the migration of 3PLs from asset-based to knowledge-based, value-added logistics service enterprises.

## 2 The role of ICT

The increasing role of information and communication technology (ICT) has contributed to the evolution of the competitive scenario in the international 3PL industry.[9] It has allowed the entry of new players in the market from unexpected industries and has led to changes in the way 3PLs conduct their business.

Three trends are evident in the impact of ICT and web technologies on the 3PL industry.[10] First, there has been increased integration of traditional services (transport and warehousing) with information services such as shipment tracking and tracing. Second, there has been widespread development of new functions for virtual intermediaries such as online freight e-marketplaces. Finally, alliances have formed between 3PLs and other companies operating in complementary sectors (i.e. ICT vendors, management consulting and financial services) that in some cases have given rise to the creation of a new category of service provider called Fourth Party Logistics Providers.

The availability of capable ICT-based services is an expected dimension of 3PL service supply. The cost of entry into the 3PL arena now includes technology and implementation capabilities for warehouse management, transportation management, and web-enabled communications. The focused efforts of 3PLs to continually upgrade and expand ICT capabilities have reduced many of the once differentiating high-value technologies to what are now minimum requirements. Users of 3PL services anticipate that the near-term differentiators will include electronic markets, supplier management systems, and supply chain planning. Going forward, the success of 3PLs will depend on their ability to deliver an integrated, end-to-end solution that provides significant improvements in financial and operational performance.

Given these trends, ICT is of critical importance to achieving the innovations needed for 3PLs to succeed in the development of new logistics services in a customised supply chain context. In this regard, Sauvage[11] noted that in a highly competitive business characterised by time compression, technological effort becomes a critical variable and a significant tool for differentiation of logistics service providers. Van Hoeck[12] assigned a specific role to ICT for 3PLs aiming to perform customising operations for service users. The use of specific technological capabilities may leverage transport and logistics services and facilitate more effective organisational and flow integration across companies in the supply chain. For 3PLs, ICT capabilities may assure the rapid customisation of products and maintain competitive lead-times. At the same time, transparency of transport and logistics operations for the customer might be important to monitor performance and assure product availability.

While today's marketplace is seeing more productive and meaningful 3PL-customer relationships evolve and customers generally report high levels of success with their 3PLs, a gap exists between what the customers receive and what they expect to receive. Consequently, 3PLs need to focus on a number of key objectives, including implementing information technologies, instituting effective management and relationship processes, integrating services and technologies globally, and delivering comprehensive solutions that create value for the users and their supply chains. Considering that customer demands for performance and sophistication are accelerating, improving these areas is a key imperative for 3PLs.

## 3 Shipping lines

Ocean carriers are no exception in this scenario as the changes outlined above have also influenced such companies, leading them to face new challenges. Undoubtedly, the ocean transport industry, particularly in the container-shipping sector,[13] has markedly changed in recent years. In the last decade, deregulation and the weakening of cartels, the huge capital investment in ships, terminals and equipment, and the development of alliances, mergers and acquisitions are all factors that have had important effects on the international liner-shipping sector.[14] As in other transport modes, the development of SCM and the dissemination of ICT and e-commerce tools in the shippers' logistics systems are major driving forces influencing liner-shipping industry. The confluence of such factors is having a significant impact on the industry. Logistics and ICT are considered dominant issues and 'seeds of change' in the industry.[15] There is a shift in the balance of power in the shipper-ocean carrier relationship essentially from carriers to shippers.

Several shipping lines have been offering freight consolidation services since the mid-1970s mostly to meet the needs of their North American and European customers that imported large volumes of goods from manufacturing centres in the Far East. But pure consolidation and deconsolidation was not enough as shippers sought more sophisticated services together with better information and control over shipments in order to minimise inventories and improve their customer service. This fuelled the trend for major liner companies to create logistics divisions, either inside or outside their organisation.[16]

This trend has picked up speed in recent years as ocean carriers have followed road hauliers, warehouse companies, transportation intermediaries and others into the logistics service market. In addition, freight forwarders and 3PLs are increasingly taking control of the cargo away from pure shipping lines. With their sophisticated ICT and e-commerce solutions and their customer service functions, these companies have challenged traditional practices in the liner shipping industry.

Under the pressures of customer demand, the largest shipping lines have extended their services from providing little-differentiated port-to-port transportation services to more customised logistics service packages. This has fuelled the gradual replacement of isolated transportation transactions with long-term supply chain partnerships based on integration of land and sea transport including port terminals and inland depots.

Though the past ten years have witnessed an unprecedented move by shipping lines into the logistics service market, inextricably linked to which has been the growth in ICT and e-commerce, the liner-shipping industry does not have a major share of the logistics service market. While most of the largest shipping lines claim to be able to provide logistics services, their impact on total business revenues remains generally low.[17]

Although little attention has been paid to the development strategies of shipping lines as they respond to the evolution of supply chain management practices,[18] different reasons can be found for this. First, shipping lines have to overcome the shippers' perception that they are not able to manage all stages of the supply chain.[19] Most container lines appear to lack the in-house expertise to provide a full range of supply chain-services and they still offer few logistics services beyond cargo consolidation.[20] It may seem that the way shipping lines conduct their business is still based on exploiting their traditional capabilities rooted in ocean transport and related services.[21] Evidence arising from a survey of alliances set up by shipping lines in the 1990s indicates that, for the most part, deals are focused on the maritime-port stage of services and do not involve inland transport and value added logistics services.[22] Second, Brooks[23] highlights the role of the perceived risks and the skill base needed to undertake supply chain operations to explain why the logistics service market is not attractive to many ocean carriers. The picture emerging indicates that shipping lines are currently undergoing a transition phase but strategies to enter the logistics service market are not very clear.

The dissemination of information technology and e-commerce has been considered an effective source of service innovation in the ocean transport sector as well as in the logistics service industry. ICT has the potential to improve co-operation between shippers, carriers and their supply chain partners. Additionally, the Internet offers the potential for shippers to benefit from real-time supply chain operations, including order/shipment initiation, en route tracking of goods, customs verification and delay alerts.

Nevertheless, in terms of the adoption of ICT and e-business systems, the liner shipping industry has strong internal EDP and EDI systems but a traditional weakness in external electronic links with customers and other supply chain participants. In addition, shipping lines seem relatively slow in implementing ICT in comparison with parcel delivery companies or large freight forwarders.[24] The liner shipping industry is notorious for poor communication among its players and mediocre information management connected with a wide range of logistical deficiencies, including:

- lack of transparent information on orders/shipments in the supply chain (e.g. shippers want to be able to track movement of their goods through the supply chain);
- workflow and procurement inefficiencies;
- poor customer service;
- reactive supply chain management;
- industry specific, information-intensive requirements (i.e. customs, legal compliance, etc.);

- poor inventory management; and
- high transaction costs and time-consuming negotiations with a shipper or a shipper's agent.

Few contributions in the literature have addressed ICT issues in liner shipping. Both the academic and practitioner literature seem to have developed along two main lines:

1  analysis of opportunities and threats connected with the dissemination of ICT and e-commerce in the liner shipping sector.[25]

   Such opportunities are generally identified as: improvement in the efficiency of the transportation services, optimisation of internal information flows, improvement in relationships with customers under a partnership approach, electronic substitution of paper documents, etc.;

2  the assessment of delay in ICT and e-commerce adoption by ocean carriers.[26] This delay is to a large extent referable to cultural factors that prevent the wide dissemination of entrepreneurial and technological innovation in the sector.

The e-business scenario in the liner shipping industry highlights the absence of a well-defined business model. Some large shipping lines have started e-commerce initiatives through the management of their own Internet portals to serve clients better through supplying on-line booking, tracking and tracing of the goods and other information and additional services (as in the case of NYK, APL, OOCL, P&O and Maersk). These portals are rarely able to give end-to-end visibility of goods along the entire supply chain.

Furthermore, the web host of a number of portals devoted to transport and logistics is managed by companies outside the shipping industry (the so-called infomediaries or e-marketplace). Some of these initiatives are 'container shipping specific', while others also operate in the air and land transport sector. Some of these portals – particularly freight auction portals – have not received great attention from the shipping lines because they mainly focus on price rather than other service elements.[27] Also, from the perspective of shippers, these new electronic channels have not achieved great success due to the anonymity that accompanies the quotations of the services that put different carriers on the same footing with others. This is confirmed by the fact that such portals often bypass freight forwarders and take no responsibility for the results of transactions.

Recently some initiatives have been launched based on joined efforts among shipping lines, logistics service providers and other companies working in complementary sectors such as banks, insurance, suppliers of equipment, etc. These initiatives have resulted in the realisation of web-shared platforms among all the companies participating in transport and logistic operations that generally aim to drive efficiencies into the ocean transportation industry by streamlining and standardising traditionally inefficient processes. The services offered by these portals allow shippers, freight forwarders, third-party logistics providers, brokers and importers to manage the booking documentation and tracking of cargo across multiple shipping lines in a single integrated process. Today, there are four main initiatives in the sector: GT Nexus, INTTRA, Cargo Smart and Bolero. The participation of ocean carriers in these initiatives is very strong. Shipping lines are also involved in e-business port initiatives (the so-called Cargo Community Systems – CCS) usually promoted by port authorities that also includes shipping agents, port and terminal operators, customs and freight forwarders.

The extent to which ICT and e-commerce help transportation and logistics operations is not clear. In 1999, the value of all logistics services in international trade was estimated at approximately $1 trillion. This included $128 billion in shipping, $196 billion in road, rail and river transport, $300 billion for port, warehousing and associated infrastructure, and $388 billion for information, transaction and associated management costs. Frankel[28] stated that by effectively using the Internet and other new technologies, carriers could reduce these costs by 50 per cent with a 15 to 25 per cent reduction in transportation costs. Only about half the total cost to a shipping company is related to ships, such as finance, personnel and fuel; the rest is administrative, sales and marketing. E-commerce could save up to 40 per cent of overall costs. Frankel also exposed the amount of wasted time and money in documentation, which he figured makes up as much as 45 per cent of the door-to-door delivery cost of products. He concluded that this 'unacceptable' waste could be largely eradicated by effective use of e-commerce and e-logistics given that it allows for much tighter real-time movement and operational control.

Summarising the above, it appears that the developments in the field of logistics and SCM on the one hand, and ICT and e-business on the other, require that ocean carriers put customer demand first instead of emphasising the operation management requirements of the container terminal operator and carriers. At the same time, there is evidence of the shift of ocean transport companies from being 'hardware-based' service suppliers towards 'know-how intensive' providers. This emphasises their search for new customised logistics solutions based on service innovation and differentiation.

The increasing importance of logistics and ICT presents shipping lines with two different alternatives: survive in a low-cost world of ocean transportation providers or pursue the innovative and problematic path of becoming value added logistics service providers.

**4 Selected works**

This part contains three chapters dealing with problems briefly summarised in the preceding pages. The chapters reflect recent changes occurring in the liner shipping industry and discuss issues that will be crucial for its future development. The first two chapters are devoted to logistics and SCM in liner shipping, while the last work addressed ICT issues in the ocean transport industry.

The first chapter (Chapter 15) entitled 'Responding to shippers' supply chain requirements' by Trevor Heaver focuses on the initiatives of shipping lines as they respond to the changing conditions. The objective is to examine how liner shipping companies are re-engineering their services to meet the changing needs of shippers and it examines the strategies of lines to extend the range of services beyond port-to-port shipping services. The chapter helps to fill the gap in the literature that has given little attention to the strategies of shipping lines in responding to the evolution of SCM practices. After discussing issues associated with the organisational relationship of vertically-related activities of liner shipping, the author describes the current state of logistics services offered by container shipping companies under common ownership with container lines. The chapter is based on a review of the current literature and interviews and communications with managers. In the concluding remarks, Heaver speculates on the future development of logistics services associated with lines. He argues that, although the logistics service industry is a growth business, risks associated with investments in such a new market remain high considering that other companies such as carriers, freight forwarders and logistics service providers are larger than shipping lines' logistics companies and well established in the 3PL market. Such conditions reveal uncertainty about benefits in terms of market share and profitability of shipping lines in the logistics business and, for these reasons, shipping lines may be reluctant to fully enter the 3PL market.

The focus of the work of Bart Kuipers' chapter (Chapter 16) entitled 'The end of the box?' examines effects of supply chain management on container operations from the perspective of the shipper. The author notes that, in today's market, innovation plays a central role in logistics and transportation sectors both in terms of service differentiation and increasing knowledge of customer logistics needs. Logistics service providers need to know what they are shipping and have to develop a proactive role related to the intentions of the shippers and customers of the shippers. According to Kuipers, the ocean transport industry presents a contrasting picture as container operations are characterised by mass-production practices and containers are considered identical boxes instead of only a way of packing a highly diversified number of goods each with differentiated logistics service requirements. Under this perspective, the author claims that liner shipping operations are to a large degree a rigid part of flexible and differentiated supply chains and he suggests some interesting options for container service differentiation and increasing flexibility in the three most important part of the trans-ocean transport chain: maritime transport, terminal operations and inland transport.

The third chapter (Chapter 17), entitled 'ICT practices in container transport' by Pietro Evangelista, analyses the impact of ICT on the liner shipping industry and shows how new technologies are gradually transforming the shipping business from a traditional hardware-based business into a knowledge-based service industry. The work first examines the importance of ICT in the logistical integration process of shipping lines. It shows that the traditional approach to business used by shipowners is unsuited to the current competitive context, where there is a growing trend of all operators to migrate towards the logistical services market. Another key feature of the work is to highlight the delay with which liner shipping companies are becoming equipped with ICT systems and web technologies in comparison with other logistics service providers such as freight forwarders and NVOCC. Furthermore, it shows that the farther electronic connections are extended along the supply chain from the shipping-port phase, the weaker they become. This puts liner shipping in a position of substantial weakness compared to other transport and logistics companies, with the real risk that they will be relegated to doing nothing more than operate vessels. In the concluding part of the chapter, the author analyses how shipping lines can take advantage of new technologies for their better integration in the customers' supply chain.

## References and notes

1  Christopher, M., 1992, *Logistics and Supply Chain Management: Strategies for Reducing Costs and Improving Services* (London: Pitman Publishing).
2  Cooper, M.C., Ellram, L.M., Gardner, J.T. and Hanks, A.M., 1997, Meshing multiple alliances, *The Journal of Business Logistics*, 18, 1, 67–89.
3  Lamming, R. (1993) *Beyond Partnership – Strategies for Innovation in Lean Supply* (Hemel Hempstead: Prentice Hall). Hines, P., 1994, *Creating World Class Supplier: Unlocking Mutual Competitive Advantage* (London: Pitman Publishing).
4  McKinnon, A., 2002, Logistics and supply chain trends, presentation given at the *Spatial Dynamics of Production, Electronic Commerce (Business-to-Business) and Transport* workshop, DLR – Institute for Transportation Research, 27–28 November, Berlin-Adlershof, Germany.
5  Kleinsorge, I.K., Shary, P.B. and Tanner, R.D., 1991, The shipper-carrier partnership: a new tool for performance evaluation, *Journal of Business Logistics*, 12, 2, 35–57.
6  Thomchick, E., Ott, J., Dornan, A. and Sisco, W., 1998, U.S. foreign freight forwarders and customs brokers: trends and customer relationships', paper given at *8th World Conference on Transport Research*, 12–17 July, Antwerpen, Belgium.
7  Daugherty, P.J., Sabath, R.E. and Rogers D.S., 1992, Competitive advantage through customer responsiveness, *The Logistics and Transportation Review*, 28, 3, 257–72.

8  Cooper, M.C., Lambert, D.M. and Pagh, J.D., 1998, What should be the transportation provider's role in supply chain management?, paper given at the *8th
World Conference on Transport Research*, 12–17 July, Antwerpen, Belgium.
9  Regan, A.C. and Song, J., 2001, An industry in transition: third party logistics in
the information age', paper given at the *Transportation Research Board, 80th
Annual Meeting*, January, Washington DC, USA.
10  Evangelista, P., 2002, Information and communication technology key factor in
logistics and freight transport, G. Ferrara and A. Morvillo (eds) *Training in
Logistics and Freight Transport Industry. The Experience of the European
Project ADAPT-FIT* (London: Ashgate Publishing Ltd).
11  Sauvage, T., 2003, The relationship between technology and logistics third-
party providers, *International Journal of Physical Distribution and Logistical
Management*, 33, 3, 236–53.
12  van Hoeck, R., 2002, Using information technology to leverage transport and
logistics service operations in the supply chain: an empirical assessment of the
interrelation between technology and operation management, *International
Journal of Information Technology and Management*, 1, 1, 115–30.
13  Bulk shipping has also seen a huge surge of interest for SCM and e-commerce
opportunities. The difference is that transactions in bulk shipping are more
complex as they are ship-based rather than container-based, Dibner, B., 2001,
Digital lessons from the maritime industry, *Mercer on Travel and Transport*,
Spring/Summer, 3–8.
14  Drewry Shipping Consultants, 1991, *Strategy and Profitability in Global Container Shipping*, Drewry Shipping Consultants, London, UK. Gardiner, P.,
1997, *The Liner Market 1997/1998 – New Alliances and The New Era*, London,
Lloyd's Shipping Economist.
15  Shirokawa, S., 2000, From maturity to growth – current state of liner shipping:
aiming for new horizon of growth, in *Current State of Liner Shipping
1999–2000*, Mitsui O.S.K. Lines Company Report, Ch 1, Research and Co-
operation Office, Tokyo, Japan (originally published in Japanese in October
2000).
16  Pontoppidan, K., 2000, The evolution in container trades and the change from
pure maritime transport to supply chain management, paper given at the
IAME 2000 Conference, *The Maritime Industry into the Millennium: the Interaction of Theory and Practice*, 13–15 September, Istituto Universitario Navale,
Naples, Italy.
17  Fossey, J., 2000, Logistics – lifestyle or lifeline, *The Drewry Container Market
Quarterly*, Section 3, September 2000.
18  Heaver, T.D., 1996, The opportunities and challenges for shipping lines in
international logistics, proceedings of the *1st World Logistics Conference*,
Ramada Hotel, London Heathrow, UK.
19  McKnight, B., Reeve, J.C. and Lee, Y., 1997, Can container lines make it as
global logistics service providers, *Transportation & Distribution*, April, 34–40.
20  Gillis, C., 1998, Shipping lines discover logistics, *American Shipper*, June, 32–6.
21  Slack, B., Comtois, C. and Sletmo, G., 1996, Shipping lines agents of change in
the port industry, *Maritime Policy & Management*, 23, 3, 289–300.
22  Evangelista, P. and Morvillo, A., 1999, Alliances in liner shipping: an instrument to gain operational efficiency or supply chain integration?, *International
Journal of Logistics: Research and Applications*, 2, 1, 21–38.
23  Brooks, M.R. and Graham, F., 2001, Maritime logistics, A.M. Brewer, K.
Button and D.A. Hensher (eds) *Handbook of Logistics and Supply Chain Management*, Elsevier Science Ltd, 419–30.
24  Drewry, 2000, *I.T. and Shipping: New Technology and New Thinking Leading
to Commercial Advantage*, Drewry Shipping Consultant, London, UK.

25 Stopford, M., 2002, E-commerce-implications, opportunities and threats for the shipping business, *International Journal of Transport Management*, 1, 55–67.
26 Bakker, S.H., van Ham, J.C. and Kuipers, B., 2001, E-shipping, proceedings of the NECTAR Conference No. 6, *European Strategies in the Globalising Markets. Transport Innovations, Competitiveness and Sustainability in the Information Age*, 16–18 May, Espoo, Finland.
27 UNCTAD, 2000, *Review of Maritime Transport*, United Nation, Geneva.
28 Frankel, E.G., 1999, The economics of total trans-ocean supply chain management, *International Journal of Maritime Economics*, Vol. 1, 1, 61–9.

# 15 Responding to shippers' supply chain requirements

*Trevor D. Heaver*

## 1 Introduction

Changes in technologies and in economic, social and political conditions combine to influence business practices and the corporate structures appropriate to execute them. The end of the twentieth century was marked by conditions that combined to result in wide attention to supply chain management as a key corporate strategy. The implications for the expectations of shippers for carriers and, in particular, for shipping lines are described in the following chapter by Bart Kupiers. The purpose of this chapter is to focus on the initiatives of shipping lines as they respond to the changing conditions. Their initiatives are largely a response to the important changes in global conditions but their actions also contribute to the evolving conditions. Changes in the practices of shippers and carriers are interactive.

The chapter reflects on-going research of the author individually and in collaboration with Pietro Evangelista and Alfonso Morvillo. The research is reflected in various works.[1,2,3] The chapter describes changes that are taking place and identifies uncertainties about the future evolution of liner and related services. Particular attention is given to the vertical structure of the industry, as the final pattern that will emerge remains uncertain.

The response of lines to the changing requirements of shippers is considered in four further sections. The next section sets out a wide framework within which the challenges to the lines and their responses can be considered. The third section of the chapter focuses on issues associated with the organisational relationship of vertically-related activities of liner shipping. The fourth section describes the current state of logistics services offered by companies under common ownership with container lines. This element of lines' strategy is still in a formative state. The last section of the chapter considers issues and uncertainties about the future development of the logistics services associated with lines.

## 2 General responses of lines to the needs of shippers

The potential responses of shipping lines to the changing conditions are summarised in Figure 15.1, a general framework appropriate for any mode of transport. This Figure places the responses into three strategic elements, although in reality the responses are multi-faceted and interactive. First, to meet the needs of shippers for access to global suppliers and markets while using a reduced number of carriers, shipping lines may increase the geographical span of their services. Second, as shippers seek improved transportation services and better integration of the components of their supply chains, carriers face issues about the range of services to provide and the level of organisational integration that is desirable among their various services. Carriers may expand the range of services offered to meet shippers' logistics and supply chain management expectations. Third, lines need to find ways to hold or reduce their cost levels. This demands new efforts to achieve economies of scale and scope, key aspects of operations in any network industry. Actions taken to pursue any one of the strategies have implications for the attainment of the corporate goals overall. However, actions are considered here separately in relation to each of the primary strategies.

The geographical span of services offered by carriers has been widened by the extension of their own services and through various forms of horizontal integration.[4,5] These include, mergers with or acquisitions of other companies and alliances or slot charters by which vessel space on routes

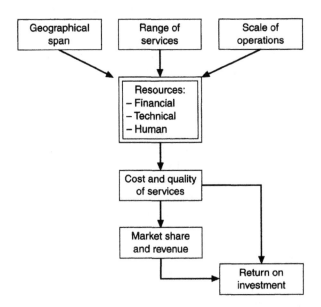

*Figure 15.1* A framework for the strategies of liner shipping companies.

can be 'shared' among lines. Through the horizontal alliances, route coverage is extended with a minimum of new investment in ships. The shift of lines to global network strategies is evident in the entry of the major east-west carriers directly into north-south routes so that the span of services offered by lines such as Maersk Sealand is truly global. Carriers that operated primarily in single ocean trades, such as CP Ships, have become global carriers, although CP Ships has followed the policy of retaining the identity of acquired lines. The growth of global carriers is associated with increased concentration in the container shipping industry. How far this concentration will go remains uncertain. It will be influenced by many factors including the economics of using larger vessels.

The extension of lines' network of services can be costly, as in other network industries, if the density of services is reduced. It is for this reason that acquisitions and slot-charter arrangements have been used so that services have been extended with limited addition of new capacity. However, new capacity has been added in the form of ever-larger ships as lines attempt to realise economies of scale by anticipating the growth of world trade. They have also reconfigured networks to realise economies of scale through the use of larger ships on long voyages to hub ports. Implementing these practices effectively is difficult. The cyclical nature of the growth in trade and the concurrence of ship orders among companies give rise to periods of depressed freight rates and low or negative returns on capital. Also, uncertainty exists about the optimal utilisation of large ships and hub ports. Concern exists that, like excessive expectations for the role of ultra-large crude oil carriers, the advantages of container vessels over 8,000 TEUs may be illusionary. The economics of vessel size is important to the configuration of services and to the potential for further mergers and acquisitions in the liner industry. The magnitude of growth in ship size over the next five years is uncertain although there is no doubt that the average size of vessels will increase.

Increasing the range of services provided by lines requires different strategies directed to the qualities of services and to their vertical relationships with other logistics services.[6,7] Investments must be made in resources to improve the characteristics of shipping services and to add service features. Improvement in information services is universal although at different levels among lines. Intermodal arrangements to facilitate door-to-door rather than port-to-port services are provided by all major lines in regions of the world where the intermodal movement of containers is practical. The extent to which lines provide warehousing and related logistics services and the extent of full supply chain management services vary among lines. The appropriate organisation structure for the range of services is evolving and is the main focus of this chapter.

The pursuit of extended shipping services, investments in new vessels and the increase in the range of services offered by lines place great pressure on the financial, technical and human resources of lines. The lines also start from different bases in terms of their size and resource mix. The companies can also be expected to have different perceptions of the costs and benefits of the strategies. It is not surprising, then, that lines follow various strategies as they strive to allocate their resources to market segments in which they can achieve appropriate costs and qualities of service to yield them profitable market shares and revenues. However, there is considerable consistency among lines in increasing the geographic scope of their services and shifting to larger vessels, even though the rate of change varies among companies. It is in adding to the range of services provided that the greatest diversity of strategies exists and the greatest uncertainty exists about the long-run outcome. The next section of the chapter considers the organisational relationship of vertically-related activities.

## 3 The organisation of vertically-related activities

Advantages can accrue to vertically integrated firms for a number of reasons.[8] They may be placed in three categories:

1  the existence of demand complementarity among businesses;
2  the presence of opportunities cost savings through the use of shared resources and expertise and through the avoidance or reduction of transaction costs among the businesses; and
3  the possibility of better profitability through increased visibility and market power and a more diversified business base.

These economics of vertical integration are considered before examining developments in the vertical structure in liner shipping.

### 3.1 Demand complementarity

Demand complementarity exists when the customer of one business is likely to become the customer of a related business. This was the reason in the nineteenth century that railways entered into the hotel business and that the airlines did so after the Second World War. The carriers had high expectations that their passengers would stay at the carriers' hotels. By operating both businesses, the carriers could take full advantage of their actions in pricing and service to attract more passengers by serving those passengers in their hotels. A decrease in the price of travel would increase demand for travel (a move down the demand curve) and would increase the demand for hotel accommodation (an upward shift in the demand curve). The carriers thus captured a part, at least, of what would otherwise be an externality, if it were other firms benefiting from the increased travel.

However, there may be offsetting diseconomies in the vertical ownership. Some demand complementarity does not necessarily mean that vertical integration is desirable. The benefits need to be strong enough to overcome disadvantages that may arise from the diversity of managerial skills and strategies that different businesses may require. The traditional separation of shipping and logistics services, associated as it has been with institutional rigidities such as shipping conferences and similar agreements, is reflected in different practices associated with pricing in shipping and logistics. In general recognition of the different skills required in different business is an important reason that firms have been changing their strategies to focus on their core businesses. Logistics is also becoming an increasingly competitive business used by sophisticated buyers. Shippers are reluctant to use a logistics service if they feel it is designed to serve as a 'feed' to another business.

In shipping and international logistics, there is potential for demand complementarity but it is not clear how strong it might be. On the positive side, the trend for shippers to prefer working with fewer suppliers increases the desirability of them being able to deal with the same firm for sequential processes. Also, there is a greater possibility for freight than for passengers that complementarity could work in either direction. However, the competitiveness of the logistics business and the potential for conflicts of interest mitigate against strong demand complementarity.

### 3.2 Opportunities for cost reduction and shared expertise

The reduction of transaction costs is a prominent reason in the economic literature for common ownership or control. Transaction costs exist in dealings between separate businesses because barriers to information flow exist as each party seeks to balance self-interest and the needs of the trade. The result is costs of research and negotiation processes and outcomes that may not be optimal. Internalisation of the transaction process can remove the barriers to information flow and can result in greater efficiency. The level of transaction costs is dependent on the actual organisation structure and practices followed. The reasons for treating owned subsidiaries as separate business units are considered later. Integrated control of businesses may enable costs to be reduced as a result of synergies. For example, related businesses may be able to share advertising costs and increase visibility by advancing a common brand name. There is also a possibility of common market and business knowledge among related services. For example, a shipping line working with shippers becomes familiar with their various service requirements and the suppliers of those services in the region. Shipping lines such as APL, Sea-Land and Maersk were able to turn such local knowledge into the basis for consolidation services in South-East Asia.

Synergies may also exist in certain cost components. Today, the most likely synergy is in information and communications technology (ICT). In supply chain management, the rapid and accurate availability of information among supply chain partners enables new approaches to supply chain management, reducing order cycle times, cutting inventories and making the systems more flexible. Consequently, the deployment of integrated ICT systems by shipping lines and associated logistics suppliers is beneficial but this does not dictate the organisational relationship between lines and logistics service providers.

### 3.3 Other benefits

The presence of a firm in various phases of a business may increase the firm's market power, for example through the better knowledge of the costs of the various operations, the availability of more service options as a result of its vertical presence or, simply, as a result of the larger size of the more integrated firm. A vertically-integrated business may also gain some benefits from diversification if the various business segments are subject to different risks, such as fluctuations in costs or to prices, and have different growth and profit potential. The latter is particularly relevant to firms able to develop and manage logistics and supply chain services.

Integrated firms providing shipping and logistics/freight forwarding services may find some disadvantages in dealing with other market participants. For example, independent freight forwarders may have a preference for dealing with shipping lines (and air lines) that do not have services competitive with them. Also, shipping lines may be less accommodating to forwarders and logistics services that are linked with other lines.

### 3.4 The experience of lines with vertical integration

Shipping lines traditionally provided port-to-port services. Since containerisation, lines have 'integrated' their activities into vertically-related businesses. The rationales and structures differ between the businesses. It is beneficial to consider the structures adopted in intermodal services and in port terminal management prior to considering the organisation structure to be in adding logistics services.

#### 3.4.1 The organisation of intermodal services

The development of the container greatly facilitated the movement of cargo between modes of transport. The ease of transfer and the greater security of cargo provided by the container led many lines to assume responsibility for door-to-door service. In effect, lines were changing the service offered by adding services previously offered by others. Shippers have remained free to make their own inland arrangements, a practice more popular in Europe than North America. The market inroad by lines as a result of better intermodal rail service has led to more competition between the lines and freight forwarders for inland transportation.

The intermodal services were not truly dependent on the container. Some companies had offered intermodal services in special situations previously. Even in the late nineteenth century, Canadian Pacific transported tea to Europe across the Pacific to Vancouver, across Canada by rail for shipment across the Atlantic. However, intermodal services were rare. Their development awaited ease and security of freight transfers.

The integration of shipping with inland services has been achieved through managers within shipping lines being responsible for the provision of inland services through a combination of owned trucking and long-term contracts and short-term purchases of trucking and other services. Shipping lines have been leaders in the development of rail-dominated intermodal services in North America, Europe and China. This is because they are in a better position than most freight forwarders to commit to the volume of traffic necessary to make dedicated rail service viable under long-term contracts. The same is now evident in barge systems in Europe. The services have resulted in great strides in rail transport quality in North America but much remains to be done in Europe and other countries.

The offer of door-to-door service by shipping lines does not raise conflicts with the interests of shippers who still have the opportunity to purchase port-to-port services and make their own arrangements for inland transport. The issue of the collective setting of inland rates is a separate matter. The entry of shipping lines into the organisation of inland transport does not give rise to anti-competitive issues. The shipping lines' services have been supported by owned trucking although its extent is limited. It gives lines a better knowledge of truck costs, greater flexibility of service level and some security of capacity at peak times. These advantages may come at the cost of somewhat higher costs. The ownership in trucking does not have any effect on the level of competition in the trucking sector because of its small size and the fragmented nature of the trucking business.

### 3.4.2 Shipping lines and container terminal operations

From the outset, Malcolm McLean believed in the assumption of responsibility for the shippers' goods and the control of as much of the transport process as possible. This concept and Sea-Land's distinctive practice of maintaining containers on chassis in terminals led Sea-Land to insist at an early date on control of it own terminals. Dedicated terminals provides lines having a sufficient volume of traffic with better opportunities than common-user terminals to integrate the schedules of their ships with terminal operations, to integrate closely ship, terminal and inland transportation arrangements and to plan and manage those arrangements consistent with the seasonality of the container business. Dedicated terminals were made available in the US in the 1970s and facilitated the development of well-integrated intermodal services. Dedicated terminals have only been made available slowly in other countries as pressures have mounted for the reorganisation of the port business.

Units of the shipping lines, for example, Maersk Ports in the case of Maersk Sealand, have traditionally managed dedicated terminals. However, the growth of container terminal management as a specialised business has resulted in the rapid growth of specialised terminal management companies. Companies such as the Port of Singapore Authority and Hutchison Port Holdings have developed global businesses. Two companies in the independent terminal-management business are subsidiaries of well-known shipping firms; they are P&O Ports owned by P&O Steam Navigation Co., which has a 50 per cent interest in P&O Nedlloyd, and Orient Overseas International Ltd (OOIL), which also owns Orient Overseas Container Line (OOCL). OOIL manages four ports through a Terminal Investment unit (the terminals are in Vancouver, New Jersey, New York and Venice) and has two container terminals operated as a part of and for OOCL. These are in Kaohsiung and Long Beach (OOIL, 1999). In the case of P&O Ports and the Terminal Investment ports of OOIL, the expansion appears to have been an opportunistic response to a new business opportunity in an area where existing knowledge could be leveraged.

The separation of lines from port management businesses minimised issues of potential conflicts of interest between a terminal operating company and its customers. However, the organisation pattern has been changed by the decision in 2001 of AP Moller to make Maersk Ports a stand-alone unit known as APM Terminals.[9] The objective of APM Terminals is 'to strive for excellence in terminal management while actively seeking new opportunities in port and terminal development'.[9] AP Moller sees its terminal management business as having a scale warranting status as a stand-alone business. It anticipates that its substantial business with Maersk Sealand (now 90 per cent of its throughput) will not jeopardise its position with other lines because of the contractual obligations between terminal operating companies and the lines they serve.

Port services provided by terminal operation are intermediate goods. The terminals do not sell major services directly to shippers. Therefore, participation in the terminal business by lines is either justified by the opportunities it provides to enhance the production of the basic transport service of the line, or it is justified as an attractive business in the service of other transport companies. In the former case, a high level of vertical integration with shipping is required. This does not give rise to conflicts with shippers interests. In the latter case, close vertical integration with the corporate shipping group could give rise to conflicts of interest with the lines to be served. It is not surprising that the specialised terminal management units are in quite separate organisation units from shipping lines but, in the case of AP Moller, that group now serves Maersk Sealand as well as other shipping companies.

*3.4.3  Shipping lines and logistics services*

Shipping lines such as APL, Sea-Land and Maersk initially entered the consolidation business in Asia to enhance their container services for domestic manufacturers and retailers. When the consolidation businesses were established they operated as separately branded and run businesses, for example, American Consolidation Services of APL, Buyers of Sea-Land and Mercantile of Maersk. This made sense as the businesses, while an extension into a related business, were quite distinct from transportation services and placed the lines in competition with freight forwarders, important customers of the lines. However, large customers in the US and Europe control a sufficient volume of traffic that consolidation services in exporting countries for a few such shippers would make the services viable. Services for smaller shippers, importing and other services in the foreign countries remained largely under the control of freight forwarders.

The development of supply chain management as an important corporate strategy has been associated with important developments in the logistics services of lines. The general growth of the third-party logistics service industry has encouraged shipping lines to expand their logistics services. Their investments have been encouraged by their belief that the logistics service business offers faster growth and better returns on capital than the lines' shipping business. Shipping lines generally now brand their logistics services with the name of the line, for example, APL Logistics and Maersk Logistics. The latest company to adopt this strategy is OOCL, which renames its Cargo System Logistics company, OOCL Logistics, effective 1 January 2003.[10] The adoption of the same brand name as the shipping line and the interests of lines in providing better-differentiated services to shippers, raise questions about the level of integration that companies are likely to achieve between their shipping and logistics services as they work to improve their services to shippers. This is considered in the next section of the chapter.

## 4  The relationship of shipping and logistics services

The growth of the logistics industry has led most transportation companies to consider how they can best serve the increased interest of shippers in third-party logistics services. Shipping is not alone in offering new logistics services. However, the organisation structure that is used and the practices that are followed differ among the modes as well as with customers. Greater independence of transportation companies and related logistics-service companies appears desirable in liner shipping than in other modes of transport. This reflects the nature of shipping services and the preferences of shippers.

## 4.1  The separability of liner and logistics services

Container shipping is not as tightly bound to other logistics activities, as may be trucking or air transport for which the design and pricing of transport and logistics services may be done in an integrated service package. In shipping, the length and uncertainty of time in transit give rise to buffer times and inventories. Shipping by a particular company remains a readily separable and substitutable part of supply chain service. In this sense, liner services are commoditised although this is not to deny the existence of some differences in service attributes among lines. Most shippers are opposed to the negotiation of combined shipping and logistics services. This is in spite of the use of, by major shippers, the consolidation services provided by shipping lines to enhance supply chain visibility and integration.

## 4.2  Most shippers retain responsibility for rate negotiation

Various factors contribute to the retention by most large shippers of responsibility for the negotiation of liner rates irrespective of who may manage logistics services. (Small shippers have long relied on rates of freight forwarders.) First, for large companies, liner contracts are very valuable and complex so that shippers may view these periodic activities as essential for them. The subsequent on-going administration of the logistics activities, cargo tracking from the time of purchase order, maintaining cargo allocations among lines and monitoring freight charges may not be strategic and may well be contracted out. Second, a shipper is not likely to see the interests of a logistics company related to a line as being sufficiently aligned with its own interest to entrust it with rate negotiation. There would be too much uncertainty about the role of the logistics company as a supplier of traffic to the sister shipping company. Third, it is likely that the negotiation of rates can be done more effectively by a large shipper than by a third party.

Shippers have intimate knowledge of the variety of competitive pressures and logistical alternatives that can be used in negotiating positions with lines. A shipper may be aware of the possibility of a new plant or a sourcing or marketing alternative that may shift traffic from one region to another, from one carrier to another or diminish the amount of ocean shipping needed in total. For example, a new chemical plant may be located close to one source material or another or close to market rather than resources. A textile manufacture may be moved from Asia to a place closer to market to achieve a short and more responsive supply chain. The pressures created by such actual or potential moves affect the ability and willingness of shippers to pay certain freight rates. While such conditions become widely known by managers in the international transport and logistics businesses, shippers can use intimate knowledge of such potential developments effectively in negotiations.

The negotiation of rate and service conditions by a shipper is usually the responsibility of a specialised management group. Such an internal organisation structure is likely to contribute to retention of the negotiation role with shippers, but it appears to do so for good reason and not just resistance to change.

### 4.3 Contrasted management attributes in transport and logistics

The separation of functions between the liner and logistics firms is associated also with contrasted attributes of managers responsible for shipping and logistics in both the service supplier and shipper firms. While the contrast may be diminishing as transport and logistics management become more integrated, differences remain significant. Sales managers in liner shipping may still communicate more effectively with the transport managers than with the logistics managers of shippers. The context of negotiations of liner rates and service conditions has changed as supply chain performance has become more important and as commercial negotiation of confidential rates become the norm, but the maintenance of leverage by shippers' transport managers is important. Shippers do this by control of traffic among routes and among modes and carriers on routes. The behavioural difference among managers is consistent with the organisational separation.

### 4.4 The corporate separation of liner shipping from logistics services

Lines are retaining the original organisational separation of liner shipping from logistics services in spite of the common branding of the organisations' names. The only company experimenting with an integrated shipping logistics group, as well as stand-alone logistics units, P&O Nedlloyd, reorganised the internal Value Added Services into the separate P&O Nedlloyd Logistics in 2002.

The separate organisation structure is consistent with the liner shipping and logistics being different and separable services offered to shippers. Potential conflicts of interests exist for shippers if the services were offered by a single business entity. Such a conflict does not exist for shippers for either intermodal services or terminal operations. Shippers also have reasons to negotiate liner rates themselves.

In spite of the consistency of the present organisation structure followed in the provision by lines of logistics services, a number of issues remain. These are considered in the final section of the chapter.

## 5  The future of lines' logistics services

The future of logistics services provided by lines and their relationship with lines are uncertain. There are unanswered matters related to the importance of logistics to the profitability of the liner shipping business.

An issue is whether it is desirable or necessary for all major lines to have an associated logistics service company. The only global carrier without such a company is CP Ships. (The line has a group called CP Logistics that is responsible for intermodal arrangements.) It could be argued that this strategy is explained by the substantial challenge faced by the line in developing its global network and the information system to serve this network and that the time has not yet arrived when investments in the logistics business are appropriate. However, it may also be that the experience of corporate management when the line was a part of CP Limited has led to a belief that diversification into related but different service delivery businesses is not a good strategy. Evergreen also was without a logistics company until 2002 when it announced that logistics services would be provided in its Asian and South American markets. This is a limited entry.

On the other hand, the strategy of Neptune Orient Line, through APL, has been to enter and expand its logistics business aggressively and to anticipate that logistics service will challenge liner shipping as a major 'breadwinner' of the corporate group. The extent to which this strategy is warranted by the expected profitability of logistics services alone or the extent to which profitable logistics is expected to enhance the profitability of shipping is not clear. Maersk Sealand has also followed an aggressive growth strategy in logistics. Both companies have made acquisitions that would enable them to provide more comprehensive logistics services involving consolidation and distribution services to shippers.

The probable effect of logistics services on the profitability of shipping is central to the logistics strategy of lines. Judging the merits of the logistics acquisitions of lines at the end of 2002 is made difficult by the global economic conditions that adversely affected the profitability of trade-related businesses. However, whatever the current situation, the outcome would only be short run. In the long run, what may be the consequences of following the logistics growth strategy? The relevant issues are the expected profitability of the logistics business itself and the effects of logistics services on the profitability of associated liner shipping.

Third-party logistics service is a growth business. As such, it is an industry with much new investment and common policies of acquisitions as companies seek to provide global services. In any industry where entry rates are high, risk exists that the expected high rates of return on capital will not be realised because of the scale of entry into the business. At least, many firms will be disappointed. It is uncertain whether the logistics companies associated with shipping lines, which are small in relation to the expanding logistics enterprises of freight forwarders, will be among the successful companies. Perhaps, in particular markets they may have a sound enough base, for example in the US, to be among the successful companies. Selecting markets carefully will be important. The size and geographical extent of the logistics services are important to the quality of service offered to shippers. Consequently, expansion into the business by firms without an established presence is not wise. This argues that shipping lines without an established base in logistics should not venture into the business. Would this put their shipping business at a disadvantage?

It has been argued previously that liner shipping and logistics services are separable businesses needed by shippers. They need to operate in a well-integrated manner with inter-operable ICT systems but common ownership is not required and some shippers see it as a disadvantage. It is likely that the logistics companies associated with lines move a greater percentage of their traffic with their associated lines than with other lines but it is not clear how far this translates into a gain in market share or what the net cost of gaining any share may be. It is unlikely that the rate chargeable on such 'tied' traffic would be higher than that on other traffic, although the business may be more reliable. It seems likely that there is some benefit in market share but whether it yields an adequate return on the investment in the logistics business is not evident from the economics of the businesses. In light of uncertain gains in market share and shipping profitability from presence in the logistics business, it is likely that some companies will choose to remain only in the shipping business.

## References and notes

1 Evangelista, P., Heaver, T.D. and Morvillo, A., 2001, Liner shipping strategies for supply chain management, *World Conference on Transport Research*, Seoul, July, 19.
2 Heaver, T.D., 2002a, The evolving roles of shipping lines in international logistics, *International Journal of Maritime Economics*, 4, 3, 210–30.
3 Heaver, T.D., 2002b, Supply chain and logistics management: implications for liner shipping, Costas Grammenos (ed.) *Maritime Economics and Business* (London, UK: Lloyds of London Press), 375–96.
4 Evangelista, P. and Morvillo, A., 2000, Cooperative strategies in international and Italian liner shipping, *International Journal of Maritime Economics*, 2, 1, 1–16.
5 Brooks, M.R., 2000, *Sea Change in Liner Shipping* (Oxford, UK: Elsevier Science).
6 Evangelista, P. and Morvillo, A., 1999, Alliances in liner shipping: an instrument to gain operational efficiency or supply chain integration?, *International Journal of Logistics: Research and Applications*, 2, 1, 21–38.
7 Heaver, T.D., 1996, The opportunities and challenges for shipping lines in international logistics, *1st World Logistics Conference*, Ramada Hotel, London Heathrow, UK.
8 Ring, P.S. and Van de Ven, A.H., 1992, Structuring cooperative relationships between organizations, *Strategic Management Journal*, 13, 7, 483–98.
9 *Economist Intelligence Unit Briefs*, 2001, Maersk Ports poised for rapid growth, 30 August, http://biz.yahoo.com/ifc/dk/news/83001-1.html.
10 American Shipper, Newswire, 4 December 2002.

# 16 The end of the box?

*Bart Kuipers*

## 1 Introduction: system innovations and paradigm shifts

The introduction of the container in the 1960s was an excellent example of
a system innovation. It was a radical innovation that fundamentally
changed the transportation industry, especially the maritime industry, and
became an important prerequisite for other revolutions like the globalisation of economic activities. During the 1970s and 1980s another system
innovation took place: the information and communication technology
revolution, having a profound influence on the maritime industry and container operations. While the container revolution originated from within
the transport industry, the information and communication (ICT) revolution came from outside the industry. The requirement of information and
communication surrounding a container is a prerequisite for successful
container management. The ICT-revolution completed the container
revolution. With ICT it became possible to make a load plan for big con-
tainerships, to manage terminal operations efficiently and effectively and
to realise intermodal transport solutions. By means of ICT terminal and
intermodal operations could react in advance of the arrival of a container
ship in the port with transport or terminal capacities. According to
Hayuth[1], the command of fast reliable information and communication is a
primary condition for entering, let alone surviving, the harsh and complex
integrated transport system. Recently, Robinson[2] suggests a new revolution – in his words: a new paradigm – with regards to the functioning of
ports and the maritime system: ports as elements in value-driven chain
systems. Here the ports are seen as elements in larger supply chains, contributing value to shippers and competing as one element in a larger
supply chain. Robinson describes the functioning of ports from a logistics
perspective and identifies that this logistics perspective itself has been
subject to development stages, resulting in the supply chain management
perspective being the current logistics paradigm. Despite the fact that
some authors claim new directions in current logistics practice like the
'demand web' approach,[3] or criticise the supply chain management
approach,[4,5] it remains the dominant logistics perspective, just like the

Just-in-time (JIT) perspective or paradigm was in the mid-1990s. In this chapter, the effects of supply chain management on container operations is described from the perspective of the dominant actor in the supply chain: the shipper. This chapter will not go as far as Molenaar[6] who predicts 'the end of the sea', but the relevance of the container as a dominant element in the structuring of supply chains will be put into question. First, the concept of supply chain management will be elaborated, in part based on the previous chapter by Heaver.

## 2 Just-in-time and supply chain management in maritime operations

Leading logistics principles and maritime transport have a difficult relation. Heaver, in Chapter 15, expects the liner shipping industry faces prospects of a diminishing role in international trade because of logistics limitations. During the early 1990s however, a distortion of the growth trend in global container transhipment volumes hardly seemed possible. In analysing factors of container growth,[7] it was argued that only in a situation in which the 'glocalisation'[8] of international business would it become the dominant economic system, this could result in a smoothing down of container growth – next to events like serious (trade) wars. The internationalisation trajectory Ruigrok and Van Tulder call glocalisation is based on the development of independent and integrated production systems in the major global regions demanding lower trade in industrial products between those regions. Glocalisation seeks to produce within trade blocs and to become relatively immune to trade barriers. In their empirical work, Ruigrok and Van Tulder[8] referred to 'Toyotism' as the prime control concept of glocalisation. Toyotism has a direct relation with the JIT production system invented and developed by Toyota.[9] The demands of tight JIT logistics principles are contrary to the supply of maritime transport – as also indicated by Heaver. JIT and maritime transport are something of a paradox (see Table 16.1) – notwithstanding the advanced use Toyota made of maritime transport under a JIT-schedule in the supply of its North American assembly facilities from Japan using concepts like merchant haulage, the use of conferences and a specific port coverage.[10] Products dependent on JIT-characteristics therefore seldom make use of deep-sea services.

A good example of logistical limitations preventing internationalisation is personal computers. It is claimed by the industry[7] that personal computers have a technological obsolescence of 1 per cent a week. Shipping a PC from Asia to Europe by sea-container service and distributing the computer to a retailer usually has a total throughput time of 10 to 12 weeks. This means that at the moment of arriving in the shop of the retailer in Europe, an Asian computer has become obsolete by 10–12 per cent. A computer manufactured within Europe is 10–12 per cent cheaper and has

*Table 16.1* The JIT-maritime transport paradox

| JIT-characteristics | Characteristics maritime transport |
|---|---|
| High speed | Slow speed |
| Short throughput time | Long throughput time |
| Small batches (<2 hours production) | Large batches usually container loads |
| Local suppliers | Global suppliers |
| Frequent deliveries | Low frequencies |
| On time deliveries | Climatic disturbances, mostly unreliable services |
| Deliveries on demand (pull) | Depending on fixed liner schedules |
| No stocks | Large pipeline stocks (several weeks) |
| Smooth flow of deliveries | Complex flow with many disturbances |
| Co-makership | Complex organisation with many parties |

the same operational characteristics, or has the same price but a 10–12 per cent improved performance. This technological rule is an important reason why European PC-firms like Tulip or Olivetti managed to survive until the mid-1990s. After that, American and Asian firms had set up European production facilities, used logistics strategies like postponement or value-added logistics (see Heaver, Chapter 15) or made use of air-transport – especially in the growing laptop segment.

Supply chain management mainly differs from JIT management in the span of co-ordination of the material flows. Heaver uses a definition of supply chain management based on Mentzer[19]:

the systematic, strategic co-ordination of the traditional business functions and the tactics across these business functions within a particular company and across businesses within the supply chain, for the purpose of improving the long term performance of the individual companies and the supply chain as a whole.

The most important difference with JIT management is the lacking of a role model like Toyota. There isn't a single and outstanding best-practice example like Toyota with regards to supply chain management. Often, JIT management is translated as the way Toyota performs its production and logistics practice. Dell Computers is sometimes suggested as such a role model for supply chain management, making use of sophisticated and innovative concepts like 'merge-in-transit'. Merge in transit is a logistics concept whereby a product on its way to a customer is assembled in a JIT fashion using components coming from geographic dispersed suppliers. These components are managed by specialised merge-in-transit software in such a way that they arrive simultaneously in a 'merge-centre'. Merge in transit can be regarded as a synchronisation of production, logistics and transport operations preventing stock of final products.[11] Merge in transit can only function by using a supply chain management approach like the one used in the definition above. But Dell Computers is something of a white elephant in the PC-industry and it is very much a question of whether Dell production and logistics practice will become the standard in the same way the Toyota principles have changed the automotive industry.

Recent research in paper supply chains – from pulp production to the printing of books, magazines and newspapers – resulted in an absence of clear supply chain management principles.[4] Also in the chemical industry,[7] almost every company talks of supply chain management – and often uses related concepts like 'business integration' as a synonym – but it is hardly possible to deduce a typical supply chain management approach in the industry. The only exception may be the way the big automotive firms use chemical firms as 'preferred suppliers' in their supply chains. The suggestion of 'improving the long term performance of the individual companies and the supply chain as a whole', in the supply chain management definition used above, seems way too optimistic for the relation between the automotive and chemical industry.

Therefore, it is somewhat disturbing that a clear definition of supply chain management and a major supply chain management role model are lacking. Supply chain management therefore simply may be translated like the use of sophisticated and innovative logistical concepts, including web-based facilities, a shared knowledge base and co-makership approaches, depending on high levels of trust between the different parties making up a supply chain.

## 3 Logistics megatrends

Four important logistical 'mega-trends' are identified in a recently performed research project aimed at identifying state-of-the-art logistics practices and trends:[12]

1    increasing logistics differentiation;
2    an increased demand for flexibility;
3    increased importance of speed in logistics operations; and
4    knowledge-management becoming a new business priority for the logistics industry.

First, there is a demand for differentiated logistical solutions between as well as within different industries. Starting in the late 1990s, all kinds of differentiated logistics practices emerged. This has implications for logistics structures, which also have shown a trend towards differentiation. At the start of the 1990s, a European distribution centre seemed a suitable general logistics solution for Asian and American firms willing to import goods to Europe. At the moment, a large number of different distribution facilities have emerged. Next to European and national distribution centres, regional distribution centres, cross-dock centres, merge centres, darkroom warehouses, European logistics service centres or rapid fulfilment depots appeared on the logistics scene. It has become difficult to design general logistics solutions. Instead, a careful match between the logistics demands of a company and tailored logistics solutions has to be designed based on important logistics characteristics like value-density, packaging-density or unique customer service requirements. The demand for differentiated services therefore asks for highly differentiated logistics solutions.

The second important trend is flexibility. The demand for unique highly differentiated products by customers has increased. In various segments – personal computers, clothing or cars – products more and more are tailored to individual customer demand. These 'mass-customisation' concepts are asking flexible production practices and accordingly flexible logistics solutions based on production to order instead of production to stock. In general, the 'customer order decoupling point', the point in the logistics chain in which production to individual customer order is separated from production to stock, is moving backwards. That means, more and more goods – and even key supplies for the final assembly of those goods – are produced on individual customer order.

Speed is a third important trend. Speed is not only required because of the differentiation and flexibility trends described above, but also because of the amount of capital employed in the goods. Containerised final products may be in transit for many weeks, like the personal computer example presented before, an undesired logistics goal from the point of view of the chief financial officer, the marketing department as well as the logistics department of a firm.

Increased knowledge intensity related to the logistics function is the fourth important trend. Logistics service providers have to know the logistics goals of their customers, related to the needs of the customers' customer. Simply transporting goods is not sufficient anymore but a carrier has to become aware of the demands of the consignee and not only the consignor. Logistics service providers more and more act as part of industrial networks and are responsible for maintaining industrial relations. They therefore have to develop sector specific production, market and logistical knowledge to become part of such an industrial network or of the supply chain as a part of this industrial network. Logistics service providers need to know what they are shipping and have to develop a proactive role related to the intentions of the shippers and customers of the shippers. They should function as industrial stakeholders instead of 'third parties' having as their main concern the managing of container operations or the load factor of container ships. If a container carrier wants to diversify in a vertical direction by offering logistics services, in the way described by Heaver in this book, their container services also should become diversified instead of using the business model of the homogenous container operation. The smaller number of logistics partners a shipper uses have a more in-depth level of customer specific knowledge and trust.

The smaller the number of logistics service providers in use, the larger the knowledge intensity or the level of trust between parties.

Like the JIT-maritime transport paradox (Table 16.1), the four logistics mega-trends described do not match very well with current container operations (see Table 16.2). Current container operations are characterised by mass-production practices: containers are considered identical boxes instead of only a way of packing a highly diversified number of goods each with differentiated logistics service requirements. Maritime container operations are to a large degree a rigid part of flexible and differentiated supply chains. The operational goals of container carriers are the prime focus in those chains. Goals related to low costs and high capacity utilisation of ships, cranes and berths. This rigid fashion is illustrated in Figure 16.1. This figure is used by a barge transport organisation to illus-

*Table 16.2* The SCM-maritime transport paradox

| SCM-characteristics[a] | Characteristics maritime transport |
|---|---|
| High differentiation for logistics services | Limited possibilities for service differentiation |
| High flexibility related to unique customer demands | Container operations are a mass market with very little possibilities for unique customer demand |
| High speed | Slow speed |
| Increased knowledge intensity and need for developing industrial relations and high levels of trust | Container operations hardly integrated in specific industrial processes or supply chains |

a  Supply chain management (SCM) characteristics derived from Kuipers *et al.*, 2002.[12]

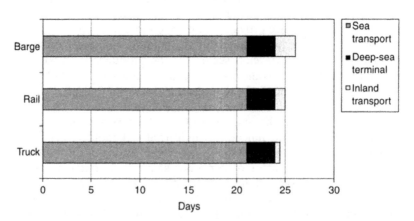

*Figure 16.1* Container travelling time Singapore-Mainz by using three different inland transport modes (source: NEA/CBRB 1995).[13]

trate the irrelevance of using inland rail transport instead of barges in a transport chain from Singapore towards Mainz, Germany, via the port of Rotterdam. The difference in total throughput time is only one day on a total transport trajectory of 24 days. It is remarkable that in the Figure only the inland part of the transport chain is differentiated and the deep-sea service and terminal operations are rigid. The example illustrates the traditional point of view related to container operations as rigid and with very little possibilities for product differentiation. This point of view does not match the current demands from logistics managers seeking differentiation, speed and flexibility. Instead of stuffing the goods in a uniform fashion in containers, containers must be adapted to the logistics diversity demanded to prevent the possible withdrawal of maritime transport operations in international trade – a possibility indicated by Heaver – in favour of local sourcing. In the maritime trajectory there are various possibilities for diversified services, but port processes remain highly rigid and 'operator driven'. In the next paragraph the possibilities for container service differentiation are presented in detail.

## 4 Differentiation of maritime services

There are various ways to ship cargo from, for example, the city of Kobe in Japan to Amsterdam, the Netherlands. Hayuth[1] presents five alternatives to ship a 10,000 pound shipment of television sets having an on-deck density of ten pounds per cubic foot (Figure 16.2) from Kobe to

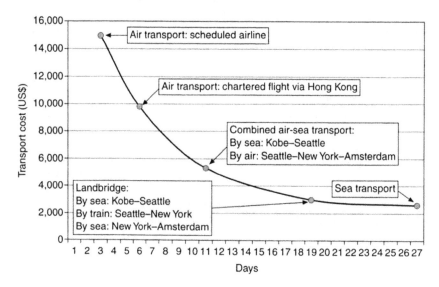

*Figure 16.2* Five different transport opportunities Kobe-Amsterdam (source: Hayuth, 1987).[1]

Amsterdam. The extreme ends of the spectrum are air and sea transport. Sea transport has a transit time of 27 days. Using a landbridge, the next alternative, has a transit time of 19 days. The time needed to transport the container from the port of Rotterdam to Amsterdam is probably included in this example. The transport opportunities between Kobe and Amsterdam however suggest, despite only five different possibilities presented, a continuous choice between the different transport opportunities. Instead of being indicated by a dot, sea transport offers a continuum between 21 and perhaps 40 days – approaching the landbridge alternative. In Figure 16.3 three different sea transport alternatives are illustrated. First, a fast trip from Kobe to Amsterdam in only 21 days: an alternative coming close to the landbridge alternative; second, a regular service, and third, a slow service. With these different sea transport alternatives a comparable all sea transport trade off between transport costs and days can be made as Hayuth presented for different transport modes combined (Figure 16.2).

The example illustrated in Figure 16.3 proves that maritime transport chains are much more flexible than assumed. In the next part of this paragraph will be illustrated how to increase flexibility in the three most important parts of the maritime chain: sea transport, transhipment and inland transport.

### 4.1 Flexibility in sea transport

The sea transport part of the maritime chain offers four possibilities to improve speed and flexibility. First, the average speed of the deep-sea container ships may be increased: ships are sailing faster and are better able to react on disturbances related to weather conditions because of more powerful engines. Hartogh[14] refers to a new type of fast container vessel

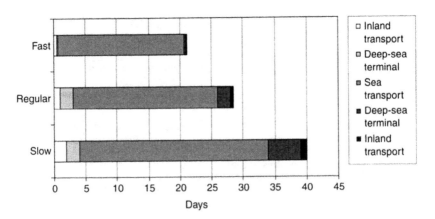

*Figure 16.3* Container travelling time Kobe-Amsterdam, using three differentiated container transport alternatives.

being developed by the Taiwanese container carrier Evergreen. These vessels will carry 4,000 TEU and will have a maximum speed of approximately 26 knots. Second, instead of making four or five calls to ports in the different regions, calls may be limited to one or two ports in a region. Third, the cargo should be loaded at the last port of call in region A and be unloaded at the first port of call in region B. This will abolish waiting time of containers aboard the container ships in other ports or hubs. Fourth, limiting the use of hubs or mainports in favour of direct calls will limit total travelling time. An important potential source of the malfunctioning of mainports relates to diseconomies of scale, especially because of congestion on the infrastructure towards the hinterland of the mainport[15] and to effects of the so called 'mainport multiplier'.[16] These multiplier effects are related to the increasing traffic density of feeder operations as container volumes and container ships will continue to grow. The same amount of containers will result in an increasing number of transhipment operations. In the airport industry, the battle between Boeing and Airbus can be related to a vision of the future position of dominant transport concepts. Will aeroplanes become even bigger than the current 747 related to more economies of scale and increased use of the hub – or mainport function – the Airbus 'A380 Superjumbo' vision? Or are customers demanding faster flights, without the annoying transfer operations and will use faster aeroplanes concentrating on direct ports of call – the Boeing 'Sonic Cruiser' vision, which at the moment however seems unlikely to be realised? Within the container industry the same discussion is taking place, between ever-larger container ships and hubbing in mainports, or so called 'Fast ships' with direct calls and initiatives like the Evergreen plans presented before. Concluding, for the high-speed and flexible demand related to the current supply chain management practice presented in the first paragraphs of this chapter – being the top-segment of the container market – relatively fast ships with direct calls seem to be preferable solution.

These four solutions together may result in a reduction of travelling time by three to five days compared to regular container services on the Asia-Europe trade, like the Kobe-Amsterdam example. Additional measures may include the use of conferences or consortia resulting in a large number of services between regions offering increased frequencies and therefore flexibility. The solutions presented have much to do with the use of container ships and port rotation schemes. But also the way containers are handled on terminals may be a source for increasing the speed and flexibility of the maritime transport chain.

### 4.2 Flexibility in transhipment operations

When a container ship arrives at a port, the operational requirements of the container stevedore are usually leading. Containers are unloaded from a container ship and are stacked at the terminal. Starting from the moment after which the container is stacked, the container may be collected by an inland transport operator for further transport towards the hinterland of the port. The time a container spends at a terminal is on average three days (see Figure 16.1). The customer order decoupling point lies at the moment in which the container is placed in a stack at the terminal, instead of the moment at which the first container can be discharged from the ship. A large container vessel, carrying 7,500 TEU, may arrive at one of, for instance, three ports on its Europe/east Asia service and may unload up to 2,000 TEU in the first port of call. In theory, it is possible that a container loaded with fashionable clothing, one week before the Christmas holidays, may be number 2,000 and will be the last container moved from the ship. With a berth performance of 150 moves per ship per hour, this will take 13 hours and may mean that the clothing will be one day later on the shelf of a retailer, risking a percentage of unsold items, for instance, because customers are shopping at other stores and buying other clothes.

At the moment, only extremely dangerous chemicals are allowed to arrive as late as possible to be the last container loaded directly from a truck, and to be the first container to be unloaded in the port of arrival. Why not offer customers service differentiation by unloading their containers as fast as possible and trying to minimise the time the container is spending on the terminal? The first 150 containers unloaded may mean an advantage of several hours. And why not allow a 'premium service' in which the container is placed directly on a truck so that the time spent on the terminal will be diminished to less than one hour?

It was stated before that, in general, customer demand becomes leading in supply chains and that the customer order decoupling point is moving backwards, indicating more production to individual customer order. In container transhipment operations, the customer order decoupling point should move backwards accordingly: from the stack towards the transhipment operation itself. This will mean additional complexity and demands to ICT in the planning of loading ships, but the rewards are important: higher prices for container carriers and terminal operators for the delivery of improved differentiated services and finally, together with the other measures described here, a higher demand for maritime services.

An important requirement for the development of the kind of services presented is the end of handling all boxes in a similar fashion on terminals. The contents of a container and the logistics characteristics surrounding this content become leading in the terminal handling operations. An important segment of the total load of a container carrier may have a first-class or business-class character, to return to air transport as an illustration. The container operations must adapt to the individual character of the container as part of an individual supply chain. Being one of the first containers transhipped, in combination with making use of fast sea-transport operations, may reduce the time spent at a container terminal, or at the berth of the container terminal, by another day. These examples of new differentiated services offering increased speed and flexibility asks new competences from both carriers and terminal operators which go beyond the traditional goals of large scale, low costs and high productivity. Container operators should become involved in the logistical characteristics of the cargo and should adopt their operations accordingly. Additional knowledge must be developed of the supply chain management perspectives instead of a container management perspective.

It is not only possibilities for the development of new maritime logistics services that ask for the development of containers as an individualised product instead of a mass-product. At the moment US intelligence services have identified 15 ships as belonging to the Al Qaeda network. The US Container Security Initiative will increase the transparency of container operations by use of container scans and other means of inspection and by demanding electronics information of the contents of containers in advance of the arrival of the container in the US. The Container Security Initiative will boost the information availability of the goods inside an individual container and therefore of the transparency of maritime transport chains.

Dedicated terminals operated by corporations having direct links to container carriers – see the examples of P&O Ports/P&O Nedloyd and Maersk ports as described by Heaver – are placed most favourably for the development of new differentiated container services. They manage sea transport, terminal operations and inland transport in an integrated fashion and have relations with individual shippers – in addition by using related logistics service providers. These dedicated terminals can easily integrate ICT-tools with the required transport and transhipment operations.

### 4.3 Flexibility in inland transport operations

Flexibility in inland transport operations was the starting point of this discussion (see Figure 16.1). It was stated that a day or more in an international transport chain is hardly a problem and therefore barge transport offers a suitable solution. This is true for a transport chain with low valued products not being a part of integrated supply chain but mainly produced on stock. In such slow moving supply chains the goods may be stored on a container terminal for several days, may be transported by sea by using a slow, low cost carrier, calling at various ports, and may be distributed by barge to an inland destination – often requiring truck transport for the final mile. But by using the supply chain as a whole, a day extra seems very relevant. With one or two days less, fast sea transport may become an alternative for sea-air transport or landbridges (Figure 16.2). With even more days less, fast sea transport might even become a competitor for air transport. In those fast supply chains, inland transport by truck, related to fast sea transport and very fast terminal operations is needed.

To evaluate different transport alternatives, the supply chain as a whole is the starting point. In the previous chapter, Heaver stated that the attractiveness of shifting to low-cost production locations has decreased recently if logistics conditions in those locations have not been compatible with the operation of manufacturing and retailing with low inventories. By offering increasingly differentiated maritime logistics services, the spatial organisation of supply chains may be influenced. The maritime part has to fit in with the required logistics demand on the level of the supply chain as a whole. This means that maritime operations – sea and inland transport and terminal operations – have to fit in with logistics requirements related to transit time and speed, frequency, reliability, flexibility and transport costs. For some supply chains a day more or less is not very critical. For others however, one day becomes very critical. But how important is one day extra from a logistics point of view? The next paragraph will give some points of departure.

## 5  Time sensitivity in the food industry

Differentiation between logistics services required by different industries and firms is one of the important trends identified earlier in this chapter. It is therefore very difficult to give a general insight into the importance of speed or flexibility for shippers. Muilerman[17] has measured time sensitivity in the food and service parts industry by using the conjoint analysis method. The service parts industry hardly makes use of sea transport, as opposed to the food industry,[18] therefore results of trade-off ratios for the Dutch food industry will be presented. Muilerman gives trade-off ratios for transit time, frequency, reliability and transport costs (Figure 16.4).

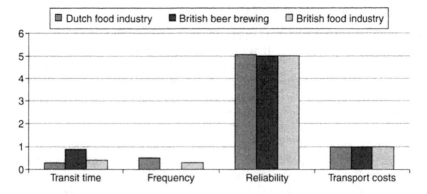

*Figure 16.4* Trade-off ratios in the Dutch and British food industry (source: Muilerman, 2001).[17]

These trade-offs indicate what the sensitivity of an average logistics manager in the Dutch food industry is for these four important logistics factors. Muilerman also presents the results of a British study related to identifying trade-off ratios for the British beer brewing industry and the British chocolate and sugar confectionery industry by Fowkes, conducted as early as 1989, matching the results in the Dutch food industry quite closely.

The ratio of 0.3 for the Dutch food industry shows that the average logistics manager in the Dutch food industry is indifferent between a change of the transit time by 1 per cent and a change of the transport costs by 0.3 per cent. Muilerman considered as the most remarkable result that, when measured in terms of costs, reliability appears to be very important. For every additional per cent reliability improvement (anywhere between 90 and 100 per cent), the average logistics manager in the Dutch food industry is prepared to pay 5 per cent higher transport costs. Ranked according to this cost ratio, reliability clearly becomes the most important attribute in the Dutch – and British – food industry, followed by transport costs, transit time and frequency respectively.

These trade-off ratios give a clue to how the average logistics manager in the Dutch food industry assesses the improvement of different logistical characteristics. Logistics managers in the Dutch food industry using inter-continental container transport will put extra emphasis on the reliability of the service instead of increasing speed and are likely to pay for an increased reliability for supply chains including container transport. Other industries however will have other trade-off ratios, in which transit time for instance will be more important.

The important point is that these kind of trade-off ratios are very important in delivering new and differentiated container and logistics services. Logistics service providers therefore have to know in detail what the trade-off ratios of their major customers – and of the customers' customer – are in the development of new logistics services.

## 6 Conclusions: customers instead of containers first

As an important player in international supply chains, container operators and related logistics service providers must adapt a new logistics strategy starting from the logistics demands of their important customers, demanding services in a supply chain management perspective. This perspective however is not very clear yet. But a number of important logistics characteristics related to current demand for logistics services seem clear. There is a demand for:

- highly differentiated logistics services;
- increasing levels of flexibility related to unique customer demand;
- high speed services; and

- an increased knowledge intensity and the need for developing industrial relations and high levels of trust in the industrial network as a
whole.

These demand characteristics are a starting point for container operators
to develop differentiated logistics solutions matching these characteristics.
The way the warehouse sector has reacted with a large number of differentiated warehouse concepts – European and national distribution centres,
regional distribution centres, cross-dock centres, merge centres, darkroom
warehouses, European logistics service centres or rapid fulfilment depots –
may act as an example.

A number of possibilities for service differentiation within the maritime
part of supply chains have been presented, offering possibilities of new
services in sea transport, in terminal operations and inland transport. It is
crucial to put the demands of the customer first instead of the operational
demands of the container terminal operator and carrier. To give the right
logistics solutions, all parties in the supply chain have to work towards the
goals and demands characterising a supply chain. If speed and reliability
are essential logistics requirements, a day sooner or later becomes a
mortal logistics sin.

### References and notes

1 Hayuth, H., 1987, *Intermodality: Concept and Practise. Structural Changes in the Ocean Freight Transport Industry* (London, UK: Lloyd's of London Press Ltd).
2 Robinson, R., 2002, Ports as elements in value-driven chain systems: the new paradigm, *Maritime Policy & Management*, 29, 3, 241–55.
3 Saabeel, W., Verduijn, T.M., Hagdorn, L. and Kumar, K., 2002, A model of virtual organization: a structure and process perspective, *Electronic Journal of Organizational Virtualness*, 4, 1, 1–17.
4 Runhaar, H., 2002, *Freight Transport: At Any Price? Effects of Transport Costs on Book and Newspaper Supply Chains in the Netherlands* (Delft: Delft University Press).
5 Cox, A., 1999, Power, value and supply chain management, *Supply Chain Management: An International Journal*, 4, 4, 167–75.
6 Molenaar, H., 2001, The end of the sea. International workshop on compara-tive Antwerp-Rotterdam port history (1870–2000), Antwerp, 10–11 May.
7 Kuipers, B., 1999, *Flexibiliteit in de Rotterdamse havenregio. Flexibiliser-ingsstrategieën van de moderne zeehavenindustrie* (Delft, the Netherlands: Eburon).
8 Ruigrok, W. and Van Tulder, R., 1995, *The Logic of International Restructuring* (London, UK and New York, USA: Routledge).
9 Womack, J.P., Jones, D.T. and Roos, D., 1990, *The Machine that Changed the World* (New York, USA: Rawson Associated).
10 Eller, D., 1992, CKD bonanza motors on, *Containerisation International*, August, 26–32.
11 O'Leary, D.E., 2000, Reengineering assembly, warehouse and billing processes for electronic commerce using 'merge-in-transit', *Information Systems Fron-tiers*, 1, 4, 379–87.

12 Kuipers, B., Becker, J.F.F., Iding, M.H.E. and Ruijgrok, C.J., 2002, *Syntheses-tudie naar trends in het goederenvervoer en innovatie* (Rotterdam: Adviesdienst Verkeer en Vervoer).
13 NEA/CBRB, 1995, *Vaart in containers. Positieschets van de containerbinnen-vaart als volwaardig intermodaal alternatief* (Rijwijk/Rotterdam: NEA/CBRB).
14 Hartogh, A.H., 2001, *International State-of-the-art in Container Logistics and Performance Requirements for Mega Hubs. A Vision for Container Logistics in the Port of Rotterdam* (Delft, the Netherlands: Connekt).
15 Klink, van., H.A., 1995, *Towards the Borderless Mainport Rotterdam. An Analysis of Functional, Spatial and Administrative Dynamics in Port Systems* (Rotterdam: Thesis Publishers and Tinbergen Institut).
16 Ashar, A., 1996, Evolution in transhipment patterns and their impact on ports. Case study: Indonesia, *Terminal Operations Conference*, 16–18 April, Hamburg, Germany.
17 Muilerman, G.J., 2001, *Time-based Logistics. An Analysis of the Relevance, Causes and Impacts* (Delft: Delft University Press).
18 Halweil, B., 2002, *Home Grown. The Case for Local Food in a Global Market*, Worldwatch Paper 163 (Danvers, MA: The Worldwatch Institute).
19 Mentzer, J. (ed.), 2001, *Supply Chain Management* (London: Sage).

# 17 ICT practices in container transport

*Pietro Evangelista*

## 1 Introduction

A growing number of manufacturers and retailers have adopted the supply chain view to manage their business in recent years. For these companies the delivery system has become an integral part of the supplied product, to the extent that transportation and logistics receive the same evaluation as the product itself. In this context, transportation providers play a more important role than in the past insofar as they are entrusted with the task of co-ordinating and, secondly, accelerating physical and information flows along multiple levels of the supply chain and of making the whole logistical system more efficient and flexible in responding to swift market changes.

Information systems and integrated transport and logistics chains are closely related since a good information management is essential for a transport company to be truly integrated. Information and Communication Technology (ICT) facilitate the management of interconnecting major information flows related to goods flows, among all actors involved in the service production process. Transportation companies have been active in developing information systems. Nevertheless, the use of ICT is irregularly distributed among the various modes. Single-mode operators such as air, railways, and road transport companies have for a long time used in-house information systems supporting their operations.

The ocean transport industry represents a good exception to the slow implementation of ICT. In the liner shipping sector, despite the efforts to open up joint information systems between shipping lines, ports and other actors in the transport and logistics chain have shown growth for the last few years, sophisticated and integrated inter-operator electronic trading has not taken off. This means that the scope ICT can offer for underpinning basic services with value-added supply chain services, has not been fully exploited by liner shipping companies.

In this chapter the impact of ICT on the liner shipping supply chain is investigated. What the work attempts to do is to bring together the Supply Chain Management (SCM) approach and perception of ICT integration in the sector. Due to the breadth of ICT and its rapidly changing nature, the work will focus on a number of key issues rather than produce a generic survey on systems and applications which would be out of date in a short time.

The chapter is structured as follows: starting point, in the second section, is the discussion of the growing important role of ICT in SCM and its implications for transport and logistics service providers. The third section discusses the impact of ICT on the liner shipping industry. In the fourth, an overview of the dissemination of and the use of ICT in liner shipping has been given. Finally, in the conclusion, issues for the better integration of liner shipping companies in the global logistics system have been discussed.

## 2 Transportation in supply chain management and the role of ICT

In the last few years, many efforts have been made by a growing number of manufacturers and retailers to balance cost savings and enhance their competitiveness with improvements in customer service. This has led such companies to move towards the adoption of the supply chain view in managing their business.

In this context, there is general agreement that effective SCM represents a fundamental tool for achieving competitive advantage in today's business environment. Cooper *et al.* (1997)[1] have defined SCM as 'an integrative philosophy to manage the total flow of a channel from earliest supplier of raw material to the ultimate customer, and beyond, including the disposal process'. The SCM approach allows addressing decisions related to supply chain in several areas such as location, production, inventory and transport both at a strategic and operational level in an integrative way.

With regard to transportation, the SCM approach has considerable implications. First, the re-engineering of physical and material flows realised by shippers affects logistics and transportation management in several ways: reducing average load size per shipment, increasing frequency, reliability and punctuality. Second, the rationalisation of the supplier networks made by large manufacturers is forcing such companies to outsource significant parts of their logistics activities, as well as select and reduce the number of logistics and transportation services providers with which to establish long-term favoured relationships. Third, Sheffi (1990)[2] stated that after the liberalisation which has occurred in the transportation markets of the main industrialised countries 'further transportation cost reductions cannot come from lowering carriers' prices but from better engineering of shippers' logistics systems'.

Such changes increasingly require a higher degree of integration of transportation and logistics companies in the supply chain environment as poor integration represents one of the main sources of chain inefficiency (Gentry, 1993).[3] The literature recognised the importance of integration in the SCM concept. Stevens (1989)[4] has described a useful model for identifying the steps needed to attain a totally integrated supply chain. This model allows for the transition from an initial phase of complete functional autonomy within the firm (baseline organisation) to a final phase in which the firm extends the level of integration achieved with suppliers and customers (externally integrating company). In this phase, gaining competitive advantage no longer depends exclusively on the level of integration within the firm itself, but rather on exploiting the advantages derived from integrating suppliers (of goods and services) and continuous improvements in quality, cost and delivery. Womack *et al.* (1990)[5] have shown that co-ordination among the various actors in the supply chain is thus an important prerequisite in order to achieve competitive advantage. Such co-ordination requires a high degree of organisational integration between the manufacturer and its suppliers of goods and services (Lamming, 1993; Hines, 1994).[6] New (1996)[7] argues that the focus on long-term relationships between trading partners and the operational integration of trading organisations in the supply chain represents the path towards a more co-ordinate and responsive supply chain.

In the SCM literature, both academics and managers have also emphasised the role of ICT as a key integration element. Arntzen *et al.* (1995)[8] claimed that the importance of ICT in SCM is demonstrated by the number of companies that have designed and implemented new information systems and technologies for SCM. ICT are meant to pervade the whole supply chain, integrating not only the functions and processes of a single company, but also those of suppliers with broad and long-term implications for an organisation's competitive advantage. To this end, Christopher (1997)[9] underlined that

> the fundamental concept is that organisations in the chain seek to create additional customer value through the exchange of information. Such a process can be referred to as the value-added exchange of information where value is created by the management of two main flows within the supply chain, namely the flow of information and the flow of materials and goods.

Pontrandolfo and Scozzi (1999)[10] noted that ICT directly affects the management of information by reducing time and cost to manage and transfer information and improve its quality (i.e. in terms of information richness, reliability, consistency, usability). This in turn makes material flows more effective and efficient. Indeed, the exchange of information frequently precedes the physical movement of materials and products, thus enabling firms to reduce inventories and use resources most effectively. Furthermore, Crowley (1998)[11] noted that due to developments in ICT, today it is easier to replace inventory with information, since information is becoming increasingly cheap and inventory is becoming increasingly costly. Through EDI, for instance, the inventory/information trade-off along the supply chain may be optimised.

It is worth noting that, while the management of information flows and availability of real-time data contributes considerably to the supply chain integration (Christopher, 1992; Johannson, 1994),[12] the capacity to supply real-time information closely depends on the availability and use of ICT in every single stage of the supply chain. As a result, poor ICT resource management by one or more companies involved in supply chain operations could have negative repercussions on the performance of the entire chain in terms of costs, planning ability and customer service (Lee, Billington, 1992).[13] For this reason, an integrated information management is not only required by shippers but also by transport and logistics service providers. Nevertheless, in the transport and logistics service industry, the use of ICT is irregularly distributed among the various modes. Vanroye and Blonk (1998)[14] identified the following factors which inhibit greater dissemination of ICT in the sector:

- the traditional resistance of transport companies to change;
- the small size of transport firms (with particular reference to the European market) which have insufficient resources to finance investments in ICT and related skills;
- the lack of user-friendly ICT tools;
- the use of proprietary standards by the most important industry players which prevents real supply chain integration.

Due to such constraints, the full potential of ICT to better integrate transport and logistics activities in the supply chain has not been fully realised. Starting from the scenario depicted above, the next section analyses the impact of ICT in the ocean container shipping industry.

## 2 The impact of ICT on ocean transport industry

As shown above, for manufacturers and retailers the widespread adoption of ICT along the supply chain represents a powerful way for integrating their business systems with customers as well as suppliers, laying greater emphasis on the linkages of such organisations through information. In addition, due to the fact that shippers are focusing on their core business, it is expected that the trend towards the outsourcing of transport and logistics activities will continue in the near future. Thus integration and co-ordination based on information management is considered increasingly necessary as shippers increasingly demand that their supply chain partners conduct the business electronically.

Under the strong pressure of customers, transport and logistics service providers have attached growing importance to ICT in the management of their business including shipping lines. These developments raise the following question: what is the impact of ICT on the liner shipping business? To answer this question it is necessary to firstly analyse whether the business model adopted by ocean carriers is consistent with changes in customers requirements and then discuss how ICT are changing the role of participants in the ocean transport industry.

Customers requirements have changed drastically in recent years, as customers demand increasingly more information and control over their shipment. Rather than simply book container shipments from one point to another, shippers have introduced more complex shipment models according to the SCM concept. Changes in manufacturing and distribution processes are increasingly regarded as a network where each suppliers serve a number of different buyers and each buyer has many different suppliers. Looking at transportation and logistics activities in terms of network has a number of implications for shippers. By keeping a certain amount at any one time, shippers can respond to buyer's request for stock much more quickly simply by taking a container out of the moving chain. They can thus create the illusion that the container was delivered from the start specifically for that customer. This obliterates the need to tie up capital in stock held in halfway warehouses. Providing this kind of arrangement requires a great deal of information technology capabilities on the part of the ocean carrier. They must be able to provide real-time information about the location of the containers, consolidated and presented in a manner which is easy for the shipper to understand. Ocean carriers are required to have co-ordinated ICT infrastructures in place which can accept last-minute destination changes and allow all of the necessary instructions to be sent to other companies involved in moving goods along the supply chain.

Generally, ocean transport operations directly involve a large number of participants such as shipper, intermediaries, modal transport providers, intermodal facility operators, port-terminal operators and consignees. Each of them perform a number of supply chain activities concerning each phase in the whole process of the flow of goods from the original shipper to the consignee.[15] In relation to the flow of goods, it is possible to identify a parallel flow of information exchanged between the parties along the chain (see Figure 17.1). Bender and Smith (1998)[16] identified two different types of information flows connected to an ocean shipment: information flows regarding the planning of transport and the information flows related to the execution of transport. Using such an approach, in the preparation phase, the process starts with an exchange of information between shipper and forwarder about price, destination, number of containers and transport time. Starting from such information, the forwarder books slots on the ship by contacting a liner agent. In this case the information exchanged consists of price, destination, transport time, number and

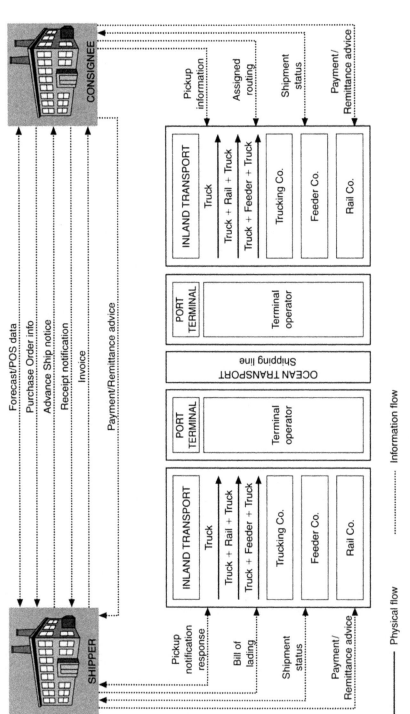

*Figure 17.1* Information flows and documents in an international shipment (source: Bender and Smith, 1998).

——— Physical flow    ............ Information flow

weight of containers. At this stage, the liner agent transmits such information to the shipping line, while the forwarder contacts inland carriers to arrange the transport of containers to the port of transhipment. Finally, the liner agent transmits booking information to the stevedore while the shipping line notifies the liner agent prior to the arrival of the ship in the port and then passes on the relevant information to other operators. The information exchanged between parties once the transport has started regards the status of the cargo, estimated time of arrival (ETA), and tracking and tracing information.

The description given above clarifies the information-intensive nature of the ocean transport industry. Considering the large number of documents produced and exchanged during an international shipment, it is expected that ocean transport operations will be increasingly driven by information technology. This requires a shift from the traditional approach to a new business model (see Figure 17.2). In the traditional perspective of the liner shipping business, each stage of a shipment is seen as an independent activity. In fact, although each participant is involved in the transportation of the same containerised cargo, conflicting objectives, interests and approaches have considerably reduced the efficiency of ocean transport operations. In such a system, many hand-overs and extensive reworking of the same information create high inefficiencies, costs and potential mistakes. For instance, the same shipment data are entered many times through the process.

Thus, efficiency optimisation is fragmented, since it is constrained within the boundaries of independent companies. Finally, in such an approach, economies of scale are considered the key competitive weapon and horizontal integration the main favoured strategic option.

By contrast, considering new customer SCM requirements, liner shipping operations have to be considered as a part of an increasingly integrated supply chain involving all parties (Evangelista, Morvillo, 2000).[17] Transport and logistics activities require tight control, co-ordination and synchronisation. To secure such characteristics and reduce inefficiencies, a

| Traditional Approach | Integrated Approach |
| --- | --- |
| • Each stage of the chain is seen as an independent activity | • Business process viewed as an integrated chain of value-adding activities |
| • Economies of scale are key competitive weapon | • Reduction of costs of both shipper and consignee |
| • Horizontal integration is the main favoured strategic option | • Vertical co-operation versus adversarial relationships with partners |
| • Efficiency optimisation is fragmented | • Reductions in direct transport costs |
| • Uncertainties in supply chain performances of other parties | • Reduction of uncertainty |
| • ICT mainly used for internal operations | • ICT used for internal operations and external integration |

*Figure 17.2* Traditional versus integrated approach in liner shipping business.

growing level of integration between parties is required. Frankel (1999)[18] argues that in the trans-ocean supply chain there is an urgent need for effective integration with customers and central control of multimodal transport and intermodal transfer or storage operations.[19] One of the reasons driving towards integration of the supply chain is the desire to have one party responsible for all door-to-door costs, with shippers and consignees assuming only delays and other costs affecting them directly. A well-integrated and effectively managed as well as controlled supply chain not only offers significant reductions in direct transport costs, but also improves the capacity and reduces the costs of both shipper and consignee. Close quality control in terms of enforcing strict schedules and transfer rates would greatly reduce costs in the liner shipping supply chain.

Another important effect related to the ICT impact on liner shipping business is the changing role and relationships among ocean carriers and other participants. The conservative nature of shipowners and the fragmentation of the business processes has been led to a situation charac-terised by poor co-operation between participants in ocean transport operations. Most of the participants, including intermediaries, are focused on a particular segment or set of activities rather than the entire process. Traditionally, ocean carriers provide transportation equipment and move containers by sea (and sometimes by land). So shipping lines have tight information control on the sea-port leg, but they remain weak on the other stages of the chain. Furthermore, other participants, such as freight for-warders, Non Vessel Operating Common Carriers (NVOCCs) and ship-ping agents, manage customer relationships usually within restricted geographical areas using their own information systems.

The increasing demand for global transport and logistic services by shippers and the developments in information and communication tech-nologies, including electronic commerce and the Internet, are changing the role and relationships between ocean carriers and other participants. For instance, freight forwarders are seeking to become truly global multimodal and logistics services providers, focusing on entire chain process rather than the narrow region of origin or destination under the traditional approach. Consequently, their key competencies and skills are shifting from traditional agency-based freight forwarding services (i.e. freight doc-umentation and customs clearance) to optimising the total transport and logistics needs of shippers. This can be successfully achieved by using information systems and telecommunications capabilities to enable them to manage and offer a comprehensive track and trace system with further supply chain management functions. A similar process is affecting shipping agents and NVOCCs.

Looking at the port industry, as ports have to better interact with their hinterland and inland transport, they are widening their role moving from a traditional interface with the ship to a more logistical orientation of their activities. This means that the landside impact on port operations has to be considered forcing ports to increase their market orientation. Consequently, this process has lead to a substantial growth of port operators on a global scale. A similar process has affected dedicated terminals. From the information technology viewpoint, this principally means extending the information flow that the port operation relies upon into a wider environment. It also brings the port's system up against other requirements – such as road or rail traffic management systems. Several major ports around the world including Rotterdam, Houston, Hong Kong and Singapore, have developed electronic port communities systems (PCS) that use Internet-based technology to connect the various parties involved in moving freight such as shippers, forwarders, insurers, customs, terminal operators, land, ocean and air carriers. A good example is Hutchison Port Holdings's e-commerce platform 'portsnportals', which offers a full range of Internet-enabled business-to-business (B2B) services through its arrangement with Arena, a leading supplier of software for SCM solutions.

The emergence of the Internet and e-commerce is further revolutionising the above scenario, enabling participants to interact quickly at low cost without following the ordered sequence in the chain. This opens the way to new types of relationships and competitive forces and can contribute to improve co-operation in the liner shipping supply chain. New types of alliances are emerging as in the case of AEI, a freight forwarder company, and P&O. They have set up a strategic alliance to send standard EDI shipping instructions and bookings and receive tracking information in return.

However, technology could also represent a threat for shipping lines as in the case of BDP International, a logistics company with great emphasis on information technology. Such a company has created an NVOCC subsidiary that can take out block bookings of container slots on ships. Once it has taken out such a block booking, which is negotiated at a favourable rate, BDP is free to sell the slots on to its own customers. The shipping line is therefore left with nothing more than owning, managing, and operating the vessels. It raises the question of who is in the best position to play the role of supply chain manager. Ocean carriers such as APL, OOCL, Maersk-SeaLand, which has invested heavily in ICT to support multimodal logistics services in addition to their basic ocean services, may be able to play such a role.

## 4 The dissemination of ICT and Internet-related service in the liner shipping industry

As shown above, the ICT impact on the ocean transport sector is deeply transforming the liner shipping business. Nevertheless, the poor communication among its players and mediocre physical and informational logistics capabilities that has characterised shipping lines for a long time still represents constraints that prevent the dissemination of technological innovation in the sector.

Although logistics service companies have used telecommunication systems and networks for some time,[20] the transport industry may not be considered a leader in the field of technological innovation (Tilanus, 1997).[21] However, over the last few years transport and logistics companies have made significant progress in the adoption of new technologies, particularly those linked to the Internet and e-commerce. Low-cost access to the Web and the dissemination of e-commerce technologies have provided these firms with the tool to satisfy customer demand by supplying traditional services in conjunction with growing information-based services. Today, the main transport and logistics service companies are in the position to provide a variety of information via the Internet[22] and to secure transactions on line with customers.

By contrast, the liner shipping industry is characterised by strong internal EDP and EDI systems (to exchange information with some agents, main stevedores, and some large ports) and a traditional weakness on external electronic links with customers. Despite the fact that for over 20 years shipping lines have been using information systems to exchange information and replace paper documents, some of these systems are unlikely to be or cannot be readily integrated into the information systems of customers or other participants in the chain. This means that uncertainties in supply chain performances of other parties (for example in terms of time and schedule risk) can negatively affect the performance of the entire supply chain.

In the last few years, shipping lines have made some progresses in the use of ICT, with particular reference to use of Internet and e-business. Such new developments have raised the question about what is the dissemination and use situation of the Internet and e-business initiatives in the liner shipping industry. Exactly how the changes connected to the widespread adoption of electronic commerce will take place and the precise changes that will come about are still much uncertain. At a first glance, the range of shipping lines on-line initiatives appears to be somewhat diversified.

Most shipping lines have initially taken the initiative to establish websites, many of which offer simple container tracking and booking services. However, this is where the initiative in most cases stops (Bakker *et al.*, 2001).[23] Today, all the largest liner shipping companies offer services through the Internet. Some of them have launched e-commerce initiatives through the management of their own Internet portals to better serve their customers (i.e. NYK, APL, OOCL, P&O, Maersk, etc.). These websites can be used not only to browse through catalogues of their services, but also to exchange information in real time, to download or upload files such as quotations, and to add and open new applications as well as the previous ones used in specific information systems through supplying on-line booking, tracking and tracing of the goods and other information and additional services. Others ocean carriers have tried to create a competitive advantage with their web pages by developing signature options unique to their brands. For example, the shipping company OOCL has developed a means to release bills of lading over the Internet. In other cases, customised portals have been developed to provide support capabilities that can also be tailored to languages other than English. APL is a good example of these advanced applications of Internet portals.[24] However, the rapid development of e-commerce is expected to give rise to a gradual increase in functionality to websites.

Nevertheless, these portals are rarely able to give visibility of goods along the entire transport and logistics chain (end-to-end visibility). In addition, there are several offerings that have not yet been made by shipping lines, although they have been made by logistics and/or parcel companies. Such developments include providing space availability information or automated confirmation of an order on line, customising the website to the level of skills of the user and connecting the user automatically to the shipping line personnel, via telephone or Internet chat, if the user runs into difficulties using the system. The main stumbling block towards creating efficient information capabilities is the lack of suitable ICT infrastructure within shipping lines themselves. Shipping lines with fragmented, proprietary computer systems, running from different mainframes in different parts of the world, will need a great deal of investment in terms of ICT before they can enable customers to handle all parts of the business on line. Shipping lines with centralised ICT systems based around personal computers communicating using Internet protocol will need much less investment to enable customers to handle business on line. Through the use of such systems, companies may be able to provide customers with the same interface that they have already developed with their personnel.

In recent years, several web portals (called infomediaries or transportation e-marketplaces) generally managed by companies out of the shipping industry have started to offer infomediary service in the container shipping industry. The potential decoupling of information flow from physical goods flow is highly significant to the future of the ocean transportation industry. While managing the information flows associated with flows of goods opens vast opportunities for new, web-based entrants to create and capture value, it also pushes the physical flow of goods – the livelihood of the various transportation intermediaries – toward merciless commoditisation. This technology-driven decoupling has the potential to exert tremendous downward pricing pressure on ocean carriers, which are currently being invited – but will eventually be forced – to co-operate with high-tech intermediaries by quoting global freight rates via the Internet. These portals are like a hub for information exchanges. Some of these initiatives are 'container shipping specific' others also operate in the air and land transport business.

Infomediaries can affect the status quo of the container shipping industry in a number of different ways. They can certainly be a timesaving tool for shippers, keeping them from having to enter many separate shipping lines and port websites in order to book and track their shipment. Shipping lines can also benefit from infomediaries, as infomediaries provide them with the access to new customer bases. However, infomediaries could also become more powerful, taking control over the booking process and the rate charged and hiding the identity of the customer from the shipping line.

Shipping lines have so far been reluctant to work closely with infomediaries, being uncomfortable about sharing their most sensitive information and unwilling to help infomediaries build up control in the market. For this reasons some of these portals (particularly freight auction portals) have not received great attention from the ocean carriers. Besides, from the side of the shippers also these new electronic channels have not received great success due to the anonymity that accompanies the quotations of the services that put different carriers on the same footing with other carriers. This situation is confirmed by the fact that often such portals bypass freight forwarders and take no responsibility for the results of transactions.

Recently some initiatives have been launched based on joined efforts among shipping lines, logistics service providers and other companies working in complementary sectors such as banks, insurance, suppliers of equipment, etc. These initiatives have resulted in the realisation of web-shared platforms among all the companies participating in transport and logistic operations that generally aim to drive efficiencies into the ocean transportation industry by streamlining and standardising traditionally inefficient processes. The services offered by these portals allow shippers, freight forwarders, third-party logistics providers, brokers and importers to manage the booking documentation and tracking of cargo across multiple shipping lines in a single integrated process. Today, there are four main initiatives in the sector: GT Nexus, INTTRA, Cargo Smart and Bolero. The participation of ocean carriers in these initiatives is very strong. For instance INTTRA carrier network includes among the others CMA CGM, Hamburg Süd, Hapag-Lloyd, Kuehne & Nagel International AG, Maersk Sealand, Mediterranean Shipping Company S.A., NYK Line, P&O Nedlloyd, Safmarine, etc.

## 5 Conclusions and implications

Reviewing issues discussed in previous sections, several conclusions can be draw. Due to the confluence of several factors, the liner shipping industry has been in dramatic transition over the past few years. Logistics plays an ever increasingly important role in such process to the extent that full logistics services are gradually replacing port-to-port transport services and isolated transportation transactions are giving way to long-term supply chain management partnerships.

In such a marketplace, liner shipping companies are increasingly asked to fully integrate the various supply chain stages through providing full sets of advanced logistics service and door-to-door solutions. Hence the need for liner companies to add value to their basic shipping services is becoming of critical importance. They have to add value to their business through increasingly supplying integrated logistics services beyond basic maritime transportation services such as inland transportation of containers, warehousing, product assembly, inventory management, etc.

A central role in the logistics integration process of ocean carriers is played by the use of ICT resources. ICT is becoming a critical attribute in any transport and logistics company to manage an integrated supply chain. It is assuming strategic importance to improve supply chain integration and to reduce chain inefficiency through supplying timely, accurate and relevant shipment information to shipper, consignee and other supply chain participants. ICT allows rationalising costs through business procedures re-engineering and, in the case of logistics services, it offers the opportunity to build closer relationships with customers through service differentiation and information system links.

A direct relationship between logistics integration and ICT co-ordination among the different stages of the chain can be found. Such relationship has been summarised in Figure 17.3 identifying three different steps characterised by different levels of logistics and information integration.[25] The Figure shows that a higher level of logistics integration requires more powerful and sophisticated information technologies. In the first stage, the

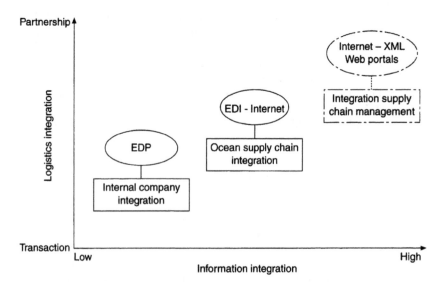

*Figure 17.3* Logistics versus information integration.

services provided by shipping lines focused on ocean services with a low level of engagement in logistics. Relationships with other transport companies and customers were based on the exchange of shipment data transmitted by traditional tools such as telephone, telex and fax. The use of information technology in this stage was limited to EDP and it was mainly finalised to reach a level of internal integration between company departments to better co-ordinate operations with administrative and accountancy needs.

Due to the increasing customers' requirement and the need for service differentiation, in the second stage shipping lines started to provide logistics services beyond pure maritime transport services. In such stage they recognised the need for a higher degree of supply chain integration. This objective can be reached through the use of more sophisticated information and communication technologies such as e-commerce tools. To make the supply chain integration process successful by using ICT tools, shipping lines have to consider two prerequisites: the interconnectivity and the interoperability. The first regards the capacity to link and get computers in communication; the second ensures that computers understand each other and properly process the exchanged information. Apart from a few companies, shipping lines have not reached a sufficient level of interconnection with both customers and other service providers. Shipping lines are well linked with port operators but the level of interconnection with other trading partners is rather low. Nevertheless, interoperability appears the area which shipping lines have to strongly improve. In the liner shipping industry information systems and tools are often based on proprietary standard. This does not allow the sharing of data and messages between shipping lines information systems with those of customers or other participants.

Finally, the third stage is characterised by the full integration driven by the massive use of ICT and e-business applications. In this stage, shipping lines are required to shift from hard-based company to software-based organisation driven by ICT. This new status is characterised by the blur of company boundaries. The control of information along the chain is necessary to the extent that the future challenge for shipping lines is to provide a continuous stream of relevant production and shipment information to customers, from factory to assembly and distribution points, through each leg of transportation, to the delivery destination and then to enable customers to quickly and easily access, analyse and act on that information.

In conclusion, the opportunities that ICT offers for underpinning basic services with value-added supply chain management has not gone unnoticed by shipping lines. Generally, ocean carriers have not obtained significant gains so far. In this scenario, the lack of systems compatibility appears the most important barrier to break down. The development of e-business and the widespread use of XML[26] standard can help shipping lines to enhance and exchange the interoperability of their information systems that is a critical pre-requisite for supply chain integration. The industry is expected to continue its gradual move forward adding more sophisticated functionality to their websites, simplifying rate databases, and improving their ICT capacity. Meanwhile, players outside the industry are moving much more quickly, creating highly sophisticated booking systems and generally making much more of an effort to look after their customers and to provide additional services including detailed information about shipment in transit, even down to a less than container load level.

The container shipping industry has reached a point where it must either abandon its traditional business approach based on cartels and information asymmetry and follow the path of innovation through offering supply chain service for their customers built on ICT and related web-technologies.

## References and notes

1 Cooper, M.C., Ellram, L.M., Gardner J.T. and Hanks, A.M., 1997, Meshing multiple alliances, *The Journal of Business Logistics*, 18, 1, 67–89.
2 Sheffi, Y., 1990, Third party logistics: present and future prospects, *Journal of Business Logistics*, 2, 2, 27–39.
3 Gentry, J.J., 1993, Strategic alliances in purchasing: transportation is the vital link, *International Journal of Purchasing and Material Management*, Summer, 10–17.
4 Stevens, G.C., 1989, Integrating the supply chain, *International Journal of Physical Distribution and Materials Management*, 19, 8, 3–8.
5 Womack, J.P., Jones, D.T. and Roos, D., 1990, *The Machine that Changed the World* (New York: Rawson Associates).
6 Lamming, R., 1993, *Beyond Partnership – Strategies for Innovation in Lean Supply*, Prentice Hall International Ltd, UK. Hines, P., 1994, *Creating World Class Supplier: Unlocking Mutual Competitive Advantage* (London: Pitman). Hines, P., 1994, Creating world class supplier: unlocking mutual competitive advantage (London: Pitman).
7 New, S.J., 1996, A framework for supply chain improvement, *International Journal of Operations and Production Management*, 16, 4.
8 Arntzen, B.C., Brown, G.C., Harrison, T.P. and Trafton, L.L., 1995, Global supply chain at digital equipment corporation, *Interfaces*, 25, 1, 69–73.
9 Christopher, M., 1997, *Marketing Logistics* (Oxford: Butterworth-Heinemann), 77.
10 Pontrandolfo, P. and Scozzi, B., 1999, Information and communication technology and supply chain management: a reasoned taxonomy, proceedings of the 4th International Symposium on Logistics, *Logistics in the Information Age*, June, Florence, Italy.
11 Crowley, A.G., 1998, Virtual logistics: transport in the marketspace, *International Journal of Physical Distribution & Logistics Management*, 28, 7, 547–74.
12 Christopher, M., 1992, *Logistics and Supply Chain Management: Strategies for Reducing Costs and Improving Services* (London: Pitman Publishing). Johannson, L., 1994, How can a TQEM approach add value to your supply chain?, *Total Quality Environmental Management*, 3, 4, 521–30.
13 Lee, H.L. and Billington, C., 1992, Managing supply chain inventory: pitfalls and opportunities, *Sloan Management Review*, 33, 3, 65–73.

14 Vanroye, K. and Blonk, W.A.G., 1998, The creation of an information highway for intermodal transport, *Maritime Policy & Management*, 25, 3, 263–8.
15 Other actors indirectly involved in shipping operations are banks, insurers, sub-contractors and bunkers.
16 Bender, M. and Smith, S., 1998, An investigation of information flows and industry analysis for the containerised shipping segment of the port of Rotterdam, Moret Ernst & Young, Rotterdam Shipping Group.
17 Evangelista, P. and Morvillo, A., 2000, Will supply chain integration be the dilemma of shipping lines in the new millennium?, proceedings of the 5th International Symposium on Logistics & the 3rd Annual Conference of the Japan Society of Logistics Systems, *Global Logistics for the New Millennium*, 12–15 July, Iwate Prefectural University, Iwate, Japan, 410–17.
18 Frankel, E.G., 1999, The economics of total trans-ocean supply chain management, *International Journal of Maritime Economics*, 1, 1, 67.
19 The author claimed that in the liner shipping supply chain more than 50 per cent of the time and costs are expended on non-transport or transport-related (loading/unloading) operations.
20 Initial applications were tried out in the air transport sector at the beginning of the 1960s. Later, their use was extended first to maritime transport and then, in the 1980s, to other transport modes.
21 Tilanus, B., 1997, *Information Systems in Logistics and Transportation* (Oxford: Elsevier Science Ltd).
22 This refers to the supply in real time of information concerning for example freight rate, booking, routing and scheduling, tracking and tracing, shipment documentation and freight billing.
23 Bakker, S.H., van Ham, J.C. and Kuipers, B., 2001, 'E-shipping', proceedings of the NECTAR Conference No. 6, European Strategies in the Globalising Markets. Transport Innovations, Competitiveness and Sustainability in the Information Age, 16–18 May, Espoo, Finland.
24 APL's Internet and e-commerce services are among the most sophisticated of this type in the liner shipping industry. Two are the most important e-services offered by the APL's website: HomePort, a customised homepage tailored specifically to certain customers, and QuickReport, an on-line service to generate automatic reports tailored to the user's specifications.
25 The concept of 'information integration' adopted here is not equivalent to the one used in computing technology. According to Teflian (2000) information integration is 'the ability to retrieve information from diverse, distributed sources, current completed and meaningful in the context of need whether for decision-making or transaction activity. Ideally, information integration could be accomplished by intuitively scouring static data repositories and dynamically integrating from multiple disparate sources into an appropriate format – virtual catalogue for instance or an on-line comparison-shopping service.'
26 XML (eXtensible Markup Language) essentially is a set of Document Type Definition (DTD) tags that carry information about specific data structures and content inside a document or file. The tags can be used by XML interpreters (i.e. web browsers) to provide a means to recognise and search for information from a wide range of databases. XML is more flexible than HTML for application-specific document markup, but lacks HTML's simplicity and pervasiveness. XML provides a standard way to define a specialised markup language, including tags specific to the application such as tcXML (Transport & Commerce XML). Whereas XML is focused on providing a mechanism to read or publish content, tcXML contains an additional set of DTD tags that support electronic commerce transactions such as cargo booking and rating and routing.

# Part V
# The regulatory framework
*Edited by Patrick Alderton and Heather Leggate*

# 18 The surge in regulation

*Patrick Alderton and Heather Leggate*

## 1 Introduction

Merchant shipping is one of the most heavily regulated industries and was amongst the first to adopt widely implemented international safety standards. Regulations concerning shipping must of necessity be developed at the global level. Because shipping is inherently international, it is vital that shipping is subject to uniform regulations on matters such as construction standards, navigational rules and standards of crew competence. It also vital to secure a level playing field for the commercial players in the maritime industry and to achieve the security of the adventure from terrorists and pirates and to prevent the adventure becoming an act of terrorism. The last generation has further seen growing concern to preservation of the environment. The regulations are wide ranging and take a number of forms such as laws, codes of practice, rules, and agreements at international, regional and national level. The regulators themselves are numerous including the International Maritime Organisation (IMO), International Labour Organisation (ILO), United Nations Conference on Trade and Development (UNCTAD), Classification Societies, the Banking Community, Unions, and pressure groups. This chapter aims to briefly survey the major areas and problems that arise from what might be considered under the general heading regulations.

## 2 The history of regulation

The Greeks in the time of Aristotle had rules similar to those governing General Average today and the Roman Merchants certainly had laws governing cargo carriage similar to those regulating Bills of Lading. In 1200 the Rules of Oleron[1] updated the Lex Rhodia which governed major areas of merchant vessel trading. For example, they governed the life and working conditions aboard ships and ensured that the Master was responsible for navigation control and cargo. Such laws were possibly codified under authority of Richard Coeur de Lion but were certainly accepted in England by the fourteenth century and in many Dutch, German and Flemish ports. However it was the industrial revolution of the eighteenth and nineteenth centuries and the upsurge in international commerce which followed that resulted in the adoption of a number of international treaties related to shipping, including safety. The subjects covered included load lines, tonnage measurement, the prevention of collisions, signalling and others. By the end of the nineteenth century suggestions had even been made for the creation of a permanent international maritime body to deal with these and future measures. The plan was not put into effect, but international co-operation continued in the twentieth century, with the adoption of still more internationally developed treaties. Co-operation at this stage was fairly simple as the number of major players was small.

By the time IMO came into existence in 1958, several important international conventions had already been developed, including the International Convention for the Safety of Life at Sea of 1948, the International Convention for the Prevention of Pollution of the Sea by Oil of 1954 and treaties dealing with load lines and the prevention of collisions at sea. IMO was made responsible for ensuring that the majority of these conventions were kept up to date. It was also given the task of developing new conventions as and when the need arose.

The history of governmental and international involvement in safety is very often the history of maritime tragedies. Governments will seldom act unless spurred on by public outrage at some recent catastrophe. The purpose of this section is to illustrate that the regulatory process is one of reaction to a problem. It is not surprising, therefore, that much modern legislation on safety can be traced back to the loss of the *Titanic*. This fact is illustrated by Table 18.1 which details such disasters and the reaction to them in terms of regulation.

The growth in media attention and public opinion has also been an important factor in driving the regulatory process. Figure 18.1 shows the

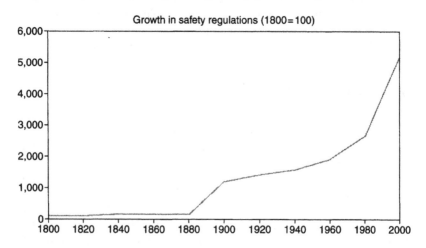

*Figure 18.1* Growth in safety regulations.

*Table 18.1* Regulation from disasters

| Year | Disaster | Year | Response |
|------|----------|------|----------|
| 1912 | Loss of *Titanic* when 1,500 people lost their lives in most tragic circumstances. | 1912 | International Ice Patrol was formed almost immediately and the irregularities in the radio watch system resolved. |
| | | 1913 | First International Safety of Life at Sea Conference (SOLAS) but its satisfactory completion was frustrated by the First World War. |
| | | 1929 | Second SOLAS. |
| | | 1948 | Third SOLAS. |
| | | 1960 | Fourth SOLAS now convened by the newly-formed IMCO (now the IMO). |
| 1967 | Stranding of *Torrey Canyon* near the Scilly Isles. | 1969 | International support for the routing of ships that had been under discussion for several years. Brussels Convention on Oil Pollution. TOVALOP and CRISTAL.[2] |
| | | 1978 | International Oil Pollution Compensation Fund (IOPC). The public *crusade* against tanker owners following the *Torrey Canyon* is a good example of how media reaction can form and create public opinion, which in turn can influence Administrations Policy. |
| 1985 | *Achille Lauro* a cruise vessel in the Mediterranean, was taken over by terrorists. One person was killed. | | It had a great effect on the Mediterranean cruise market for a decade and caused the passenger ship industry to take stock concerning its security measures. |
| 1987 | *Herald of Free Enterprise,* a tragic loss of a cross-channel ferry (193 deaths) which capsized just after leaving port with its bow doors still open. | | This caused most Ro/Ro ferry operators to improve their safety procedures and to ensure that bow doors are closed/opened on the berth. It also caused the IMO to take a hard look at the potentially unsafe features in the design of this ship type. |

*continued*

*Table 18.1* Continued

| Year | Disaster | Year | Response |
|------|----------|------|----------|
| 1989 | *Exxon Valdez* oil spill. | 1990 | US Administration introduced OPA 90 which greatly increased shipowners liability concerning oil spills when trading to the US. These regulations came into force in 1995 and ships trading to the US require Certificates of Financial Responsibility. |
| 1990 | *Scandanavian Star.* | | Raised the problem again of communication in emergencies between crews operating in different languages. |
| 1994 | *Estonia* (900 deaths). | 1998 | IMO introduce the International Safety Management (ISM) Code which became mandatory for passenger ships, tankers and bulk carriers over 500gt on 1 July. On 1 August the International Convention of Standards of Training, Certification & Watchkeeping for Seafarers (STCW) came into force. |
| 2001 | September 11 terrorist attack on World Trade Center in New York. | 2002 | Calls for wide-ranging security measures to counter terrorist threats. In May 2002 IMO has proposed 12 measures to bolster ship and port security. |
| 2002 | VLCC *Limburg* damaged by terrorist bomb(boat) off the coast of Yemen. | | This increased the pressure on ports to review their security. Also at this time, following the Estonia disaster, the IMO floated a new protocol which will come into force 12 months after it is ratified by ten states. This protocol makes far reaching reforms to the 1974 Athens Convention covering the carriage of passengers and their luggage. |

growth in safety regulations since 1800. The first rise in the 1880s coincides with large circulation newspapers and the larger rise in the 1960s to the almost universal coverage of world news by television.

Ma[3] argues in his paper that the regulators have moved in a cyclical manner in many of their attitudes in the last three decades:

As far as commercial maritime policies are concerned, if the 1960s and 1970s are characterised as the years when many countries adopted interventionism to directly or indirectly support their national shipping industry, the 1980s and 1990s can be labelled as the time when the general policy trend was deregulation and liberalisation. National shipping companies were either dissolved or privatised, cargo reservation and operation subsidy practices were either abandoned or minimised. Currently, there are some new changes, which indicate a swing back in the maritime policy trends. There are clear signs that governments are targeting shipping again and interfering in various ways in the commercial, technical and operational aspects of maritime transport. The impact of economic Globalisation, the pressure of public opinion on safety and environment and most of all the terrorist attacks, especially that of September 11, have all contributed to the change in attitude of many governments towards shipping.

## 3 The regulations

The regulations may be classified according their sphere of interest. These are basically the ship, the ship operation, the cargo, persons on board the ship, and the environment.

### 3.1 The ship

The regulation of *the Ship* is very much the domain of IMO and International Conventions. **SOLAS** (International Convention for the Safety of Life at Sea, 1974) lays down a comprehensive range of minimum standards for the safe construction of ships and the basic safety equipment (e.g. fire protection, navigation, lifesaving and radio) to be carried on board. SOLAS also requires regular ship surveys and the issue by flag states of certificates of compliance. **COLREG** (Convention on the International Regulations for Preventing Collisions at Sea, 1972) lays down the basic 'rules of the road', such as rights of way and actions to avoid collisions. **LOADLINE** (International Convention on Loadlines, 1966) sets the minimum permissible free board, according to the season of the year and the ship's trading pattern. In addition to these, the Classification Societies have vast and complex rules to ensure that ships are safely designed, built and maintained. It is true that for many flags these rules are not mandatory but there are considerable commercial pressures to ensure that Owners keep their vessels safe.

### 3.2 Ship operation

**The International MultiModal Transport Association (IMMTA)**, a world-wide trade and transport non-governmental association working in the interest of efficient trade and transport, has decided to respond positively to the global concern regarding the trustworthiness of companies involved in trade and transport. From 1 January 2003, IMMTA members may apply for certification as an IMMTA approved MTO (multimodal transport operator). IMMTA has taken this step in order to assist MTOs that may not otherwise be considered reliable or trustworthy partners in today's transport world. Three levels of certification are offered, Silver, Gold and Platinum, each corresponding to a certain number of predetermined benchmarks. While IMMTA cannot certify the performance of its members, the intention is to allow governments, shippers, consignees, banks and insurers to know that an IMMTA approved MTO complies with a certain level of minimum benchmarks. Conferences, Consortia & Alliances in the liner trade are carefully monitored by many national administrations. The Federal Maritime Commission (FMC) – the main US body concerned with monitoring the activities of liner companies has long taken strict views but these have been modified by the Ocean Shipping Reform Act (OSRA) 1998. The European Union have tended to give exemption to most liner agreement practices though like other administrations has worries about liner agreements concerning inland transport. Canada has the Shipping Conferences Exemption Act of 1987 and Australia has its Trade Practices Act of 1974. The regulatory bodies above will no doubt be concerned about the growing power of the top-20 liner carriers which controlled 72 per cent of world slot capacity in 2000.

The last few years has seen a trend in sharing responsibility with the shipowner and Management as with the **ISM** (The International Safety Management Code, 1993) which effectively requires shipping companies to have a licence to operate. Companies and their ships must undergo regular audits to ensure that a safety management system is in place, including adequate procedures and lines of communication between ships and their managers ashore.

### 3.3 The cargo

Dangerous cargoes are carefully regulated, and as crude oil is one such cargo these regulations effect a big percentage of cargo moved by sea. There is similar legislation concerning the problems in the carriage of dry bulk cargoes, specifically grain cargoes and deck cargoes. Cargo equipment is governed by the factory acts, **ISO** (International Standards Organisation) and other agencies which lay down rules concerning size and safety of containers and other unit load devices

The major rules concerning cargoes however, relate to damage and loss of cargo. The regulators governing this area are mainly insurance companies and P&I Clubs (Protection and Indemnity) whose interests are governed by the Laws governing the Carriage of Cargo By Sea, e.g. the Hague Rules, Hague Visby Rules and the Hamburg Rules.

### 3.4 Persons on board ship

Uniform standards of competence for seafarers were established by IMO in the form of **STCW** Conventions (International Convention on Standards of Training, Certification and Watchkeeping for Seafarers, 1978/1995). These set out minimum requirements not only for the qualifications themselves but also the institutions providing the training.

Acceptable standards and conditions of employment come under the auspices of the International Labour Organisation (**ILO**). Established in 1919, it brings together governments, employers and trade unions in order to facilitate this process. To this end it has adopted more than 50 conventions and recommendations covering:

- minimum age;
- compulsory medical examination prior to employment;
- no agency fee for shipboard employment;
- repatriation – seafarers must be repatriated at end of employment contract;
- entitlement to social security benefits;
- particular minimum standards for food, catering and accommodation;
- standards for vocational training and welfare facilities; and
- measures to prevent occupational accidents on board, and investigation procedures for such accidents.

Member states have an obligation to provide regular updates on measures taken to implement the conventions. There is an established system of supervision whereby such reports together with comprehensive analysis of national legislation and provision of collective agreements are examined each year by the Committee of Experts on the Application of Conventions and Recommendations. Comments from the committee are communicated to the relevant member states.

A recent example in this area is the new ILO convention which establishes a universal seafarer's identity card which has been welcomed by the International Transport Workers Federation (ITF).

### 3.5 Environmental regulations

Maritime transport is almost certainly the most environmentally friendly of all transport modes. All the evidence indicates that over 99.98 per cent of all oil carried by sea reaches its destination without incident and that tanker casualty rates compare more than favourably with those recorded for other modes of transport. In spite of these positive statistics, there is increasing worldwide intolerance of incidents involving ships because of their impact in terms of environmental pollution. Pollution is therefore the focus of environmental legislation in the maritime sphere.

**MARPOL** (International Convention for the Prevention of Pollution from Ships, 1973/1978) contains requirements to prevent pollution that may be caused both accidentally and in the course of routine operations. MARPOL concerns the prevention of pollution from oil, bulk chemicals, dangerous goods, sewage, garbage and atmospheric pollution, and includes provisions such as those which require oil tankers to have double hulls.

Other environmental impacts include vessel emissions, pollutants drained from cargo terminals, dredging for vessel access, and marine growth on vessel hulls. There is also a threat presented by the waterborne mode associated with disturbing the balance of localised marine ecosystems. Over the centuries the accidental transport of harmful insects, vermin and disease has been guarded against to protect a region's agriculture or the health of its population. What was taking place in the waterways went largely unnoticed. Now with vessels that carry large quantities of ballast water for a variety of operational purposes, the transport of unwanted marine organisms from one region to another has become a problem.

### 3.6 Security

Security issues relate to pirate activity and more recently the threat of terrorism following the events of September 11 2001. Between 1991 to 1995 over 500 incidents of piracy or armed robbery on ships and seafarers were recorded by the IMO. In 1992, the International Maritime Bureau set up a piracy centre. Pirate attacks on ships increased by 40 per cent in 1999 over 1998 (202 to 258) according to the IMB report issued in January 2000 (469 in 2000 and 335 in 2001). This report indicated that a typical pirate attack took the form of around ten men in a speedboat trailing and boarding a cargo ship. These attacks were usually violent and some indicated that the pirates knew what they were looking for and which containers to plunder. Lloyds List in 2002 suggests that the annual bill for Piracy now exceeds $25 billion. However there are few regulations concerning this problem as there is no international police force.

The problems caused by security will almost always involve considerable delays and expense with seldom the possibility of any increase in revenue and few customers will be enthusiastic to pay more for what seems to be a less efficient service. In 2003, a consultancy paper reported in Lloyds List suggested that the security costs per standard container was probably between $5 and $10. As a result of security concerns IMO has developed the **International Ship and Port Facility Security Code (ISPS)** as an amendment to come into force in July 2004. Its 85 pages will include the provision of security officers on board ships, ports will have to complete a Port Facility Security Assessment and contracting governments will have further responsibilities.

## 4 The problems of enforcement

As with any form of regulative process there is inevitably the problem of enforcement, a problem which the creators of regulations often tend to ignore. Even serious problems as ships pumping oil into a busy shipping lane in the dark is difficult to police and who pays to provide policing in international waters? Before say 1970, the situation was easier as the majority of the world's shipping was controlled by a few nations with experience in the business. However, following the very high freight rates in 1970 many speculators became interested in the business and many flags became involved that lacked the regulatory knowledge. This problem was compounded by the oil price in 1973 which saw the freight rates fall and caused an excess of supply over demand, a situation which has more or less prevailed ever since. The subsequent lowering of freight rates in real terms meant that virtually all ship operators have had to exercise their talents in cost-cutting operations and some will be tempted to reduce expenditure on non-revenue earning and costly regulations. In attempts to catch the defaulters those in authority are goaded by the media to produce more regulations until the industry is in danger of being swamped by myriads of rules, some of which are confusing, some absurd, some conflicting and some impossible to comply with. For instance, when it was first ruled that oily washings could not be dumped into the sea, captains were faced with a dilemma by charter parties which instructed them to have clean tanks to load into but the port provided no facilities for them to off-load their sludge.

Thus a major practical problem of safety is not the lack of adequate legislation but the enforcement of existing legislation. Traditionally this has been done by the flag state – but not all states have the will or the trained staff to enforce the legal safety requirements. To overcome this problem the 1978 SOLAS Protocol extends this enforcement to Port states. In 1982, 14 West European nations signed the Paris Memorandum as a bid to operate Port State enforcement in a uniform manner on foreign ships. Details of Port State inspections are kept on the Computer Centre which is located at Saint-Malo, France.

IACS has a unique technical voice and influence in the progress of world maritime safety, regulation and the prevention of marine pollution. With an unrivalled knowledge of the world fleet, the 11 IACS members and two associates classify more than 90 per cent of merchant tonnage – over 450m grt. Together, the IACS societies invest over $70 million annually in ship design and safety research. However the question often raised in the last decade is, how does a ship kept in full class with its Classification Society have bits fall off it. This does not seem a lack of policing by the Society but the difficulties facing a small team of surveyors trying to inspect vast holds and tanks in poor light and in difficult conditions.

## 5 Selected works

Rowlinson and Wixey in 'Green shipping: European policy and economics forces' (see Chapter 19) explore the conflict between safety and environment and efficiency and competition. They suggest that European policy can achieve a sustainable transport mode in intra-European freight flows without compromising safety or environmental concerns.

Safety and efficiency is threatened by a number of global maritime issues, not least the question of flag standards posed by open registry fleets, and developing nation fleets, and the emergence of new European nations. The risks and possible consequences of maritime accidents need to be minimised. It needs the involvement of all actors from all levels of decision making to accept the challenge so that effective combinations of strategies can be adopted in particular situations to address the key problems identified. However, despite maritime disasters, the potential of shipping as a substitute for more environmentally damaging or costly transport modes is clear. It is necessary to recognise the economic forces at work which could negate the environmental benefits of a switch from roads to water, if maritime standards are diluted by the need to reduce costs. Following major tanker disasters such as the *Erika* and *Prestige,* European policymakers will face major challenges in attempting to balance the economic-environmental pressures.

Li and Wonham[4] analyse the impact of legislation and regulation through an examination of occupational accidents. These account for approximately 90 per cent of all accidents at sea, much more than vessel casualties. They argue that IMO claims to focus on the 'safety of life at sea' but in fact concerns itself almost exclusively on the safety of the vessel. Further, the implementation of regulations covering life at sea is inadequate and there is a need for IMO and ILO to co-operate in order to strengthen this area and reduce the number of fatalities.

William O' Neil[5] stressed the huge amount of expertise drawn on by IMO in the production of their standards, guidelines and codes of practice. The scrutiny to which they are subjected is a particular strength of the organisation. In matters of safety, environmental protection and security the industry look to IMO to create and raise the standards. Indeed, under IMO standards have reached unprecedented levels.

It is unfortunate that shipping has such a poor public image. A few bad operators are to blame for this, eclipsing the vast majority of shipowners who fulfil their obligations and are committed to high standards. The media further exacerbate this perception problem and politicians react with increasing numbers of regulations. More regulation or even higher regulation is frequently not the best solution.

> All the regulation in the world will not prevent accidents from occurring, and so we must address the need to pursue sound maintenance and operational practices, if casualty rates are to be improved still further. The regulatory regime definitely has its part to play, but we should be clear, the reason that ships fall apart, flounder, collide with each other, run aground, break up, catch fire, or whatever else may be fallen, is rarely if ever, because there is something fundamentally wrong with the regime. In the vast majority of cases it is because somebody, somewhere along the line, did not take a proper action to avert a problem, or did something wrong, whether through laziness or ignorance, greed, malice, fatigue, negligence, whatever the fact may be, which incidentally stresses how vital it is that IMO's focus on people, as demonstrated by the introduction of the International Safety Management Code, and the updating of the Training Convention must be sustained.[6]

He further addressed the issue of enforcement. If dramatic improvements are to be achieved, then IMO needs the authority to ensure that flag states are properly implementing these regulations. IMO is currently examining such a system.

At a seminar in Bergen in the late 1990s on 'How to regulate Shipping in the Future' the chairman of IACS stated that 'It is important to share this understanding that "regulation" must be seen as a positive force to guide the future health and responsibility in our industry. All the safety partners are working towards the same ultimate goals of building, operating and regulating a world fleet to ever higher standards of accountability.' It is to be hoped that these fine and simple sentiments will achieve all that the speaker anticipated.

## References and notes

1 Oleron is a small island off La Rochelle, then English territory and an important trading centre.
2 Shipowners realising that there would be public demand for punitive legislation to prevent oil pollution, tried to remove the necessity for it by instigating a scheme known as TOVALOP (Tanker Owner's Voluntary Association for the Liability for Oil Pollution). Under this scheme tanker owners would volunteer to clean up any pollution they caused. The oil industry also offered to increase the money available for this procedure by a scheme known as CRISTAL. These offers however, were made too late, in that the mechanism for severe anti-pollution laws had been set in motion. TOVALOP and CRISTAL ended on February 20th 1997.

3 Ma, S., 2003, The swing of maritime policy trends, International Maritime Policy Conference, May, London.
4 Li, K.X. and Wonham, J., 2001, Maritime legislation: new areas for safety of life at sea, *Maritime Policy & Management*, 28, 3, 225–34.
5 O' Neil, 2004, Raising world maritime standards, *Maritime Policy & Management*, 31, 1, 83–6.
6 Ibid.

# 19 Green shipping

## European policy and economics forces

*Mervyn Rowlinson and Sarah Wixey*

## 1 Introduction

Europe's vision of a safe, environmentally acceptable shipping industry was outlined in the recent EU White Paper 'A Time to Decide'[1] and is much at the forefront of European maritime strategy. The White Paper highlighted shipping as an environmentally acceptable alternative to road haulage as it is perceived as being fuel efficient, produces comparatively low carbon emissions and can increasingly be used as a substitution for trucks, thereby leading to a significant reduction in congestion, road accidents and road damage. Against this vision is the implicit need to attain an efficient and competitive open shipping market that serves the economic needs of European trade. In order for shipping to be competitive with road, costs have to be carefully managed. There is some evidence to suggest that this can only be achieved at the expense of safety considerations (see below). The general tensions between these two positions shall provide the focus for this chapter

The increased media and political concern towards the marine environment following a number of tanker accidents over the last three decades, including the more recent 1999 *Erika* and 2002 *Prestige* disasters, has focused attention on the quality of vessels trading and transiting the community's coastline. The challenge to the maritime policymakers of the community is how to facilitate areas of growth whilst simultaneously improving safe and environmental standards of operation, particularly within the context of the expanded European Union.[2] Three market areas have been identified here as providing considerable opportunities for the shipping industry. These are:

1  the coastal and shortsea trades;
2  the Baltic oil trades; and
3  the Caspian/Black sea oil trades.

This chapter sets out to identify and discuss some of the principal challenges to a safe and quality oriented merchant shipping industry in European waters. The chapter begins by outlining the current situation and introduces the dichotomy between policy objectives and economic realities. It goes on to explain the development of the 'European Maritime Policy', particularly within the context of an extended European Community. This is followed by a closer examination of the coastal and shortsea trades, in particular the Baltic, Caspian and Black Sea trades.

## 2 The problem outlined

Seaports in the Union handle more than 90 per cent of the Union's trade with countries and approximately 30 per cent of intra-EU traffic (freight: 2.7 billion tonnes per annum). This fact was highlighted in the Commission's Green Paper on Seaports and Maritime Infrastructure,[3] in the proposal to integrate ports and terminals into the Trans-European Network Guidelines[4] and in the Report on the Implementation of the TEN-T Guidelines.[5] In the words of Loyola de Palacio, Vice President of the Commission:

> Led by economic growth up to 2010, goods traffic should increase by 38 per cent. In the long term, the economic competitiveness of the Union will be compromised, pressure on the environment and congestion will reduce safety and the quality of life.[6]

Shortsea shipping represents 41 per cent of the goods transport market, compared to 44 per cent for road. If the current situation continues between now and 2010, the Commission predicts that road goods vehicle traffic alone will increase by nearly 50 per cent.[7] The EU White Paper recommends that the Community support its rail, maritime transport and inland waterways to allow the market shares of each mode to return to the level of 1998 by 2010 in order to achieve a re-balancing between the modes.[8] The oil trades of the Baltic and the Caspian/Black Sea are set to grow significantly. For example, new pipelines from the Caspian to Black Sea and Mediterranean tanker terminal will increase trade by around 32 million tonnes per year, an equivalent of over 400 extra Aframax loadings.[9] Given the extent of this growth, the dichotomy between policy objectives and economic realities will become clearly evident in the selected issues of European shipping and trade, as identified in this paper.

In 1995, Aspinwall identified the growing cleavages between economic groupings within European shipping – the shipping companies, the shippers and maritime labour.[10] The *Erika* and *Prestige* tanker disasters have highlighted these cleavages and, moreover, have induced a further interest grouping. This is composed of the 'at risk' coastal states, particularly the Western Atlantic states of France, Portugal and Spain and the Scandinavian states, Denmark, Norway, Sweden and Finland. Essentially a twin-track approach is emerging whereby the high standards of Northern European policies appear to be at odds with the economic focus on reduced costs, as exemplified by globally footloose ownership and management of shipping and trade.

The impact of the dual effects of increased oil movements and political changes in the extended European Community could prove detrimental to attempts at improving maritime standards. At a time of intense scrutiny of the shipping industry, the political process is about to be challenged by the economic forces of the market. For Russia and the other Baltic States of the former Soviet Union (FSU), the export of oil is a major catalyst for restoring economic growth. Following the economic downturn, post-1990, output fell in most of the FSU's by up to 40 per cent; exploiting oil reserves for the export trades is seen as an economic lifeline, certainly for Russia's President Putin.[11]

Essentially, the problem is that of balancing the political demands for a safe, environmentally sustainable shipping supply chain with a globally focused shipping industry and its cost sensitive customers. A further intensification of the problem lies in the vast gulf that exists in standards of maritime safety in the market. Generally, it can be stated that vessels operated under the national registry flags have higher safety standards to those of open registries.

The movements towards such globally recognised and practised standards such as the Standard of Training Certification and Watchkeeping (STCW) and the International Safety Management (ISM) Code can be seen as narrowing the gap in safety, Lord Donaldson has succinctly summarised the positive impact of the ISM Code:

> It will concentrate the minds of shipowners who accept ship safety and ship management as desirable objectives, but never seem to be able to get round to it. It will tend to lower the cost differential between the best shipowners and others.[12]

In addition, the Paris Memorandum on Port State Control (PSC) has enhanced port inspections of vessels in European ports.

Despite these positive moves to create a better regulated, managed shipping regime, the conflicts identified here juxtapose a qualitative move towards improved standards against those of a quantitative based, economically driven need to achieve lower costs in the competitive market.

A number of critical safety issues have manifest in the selected three areas of trade. These issues are identified as:

- safety standards in shortsea and coastal shipping;
- the age and condition of tankers in the expanding trades of the Baltic; and
- the Caspian/Black Sea oil flows.

The importance of shortsea and coastal shipping has been clearly identified in European transport strategy. However, the decline in the number and aggregate tonnage of ships directly owned under national flags raises questions over standards. Quality national flag operators face intense competition from within the shipping market and, moreover, the road haulage sector. In addition, the debate on tankers has also divided along lines of coastal states anxious over the vulnerability of their beaches and the shipowning/oil trading states. The contrast in standards is evident when analysing the differences between trading vessels. For example, comparisons can be made between two crude oil loading ports: Hound Point in the Forth Estuary, Scotland and the Ceyhan port in Turkey, which is a major outlet for Caspian Oil. Figure 19.1 shows a sample of tanker movements from these two ports in the period between 21 December 2002 to 20 January 2003. There were 19 and 21 tanker movements in the ports of Hound Point and Ceyhan, respectively.

The Figure shows that the Scottish terminal has a much lower average vessel age of 5.9 years, whilst the Ceyhan terminal averages 12.3 years. During the time period, six vessels aged over 20 years were loaded in Ceyhan; at Hound Point the oldest vessel was 16 years. The evidence of tanker movements between these two ports clearly shows that a more cautious stance towards safety standards exists within the Scottish port. It could be argued that the reason for the higher standards at the Hound Point Terminal is due to the intense environmental scrutiny that its operators, BP, experience in the Firth of Forth region.

Another contrast exists in the schism between the coastal states and the shipowning/trading states. The reaction, post the 2002 *Prestige* disaster, has accelerated the pressure on withdrawal of single hull tankers. However, for Greek owners the acceleration could have a drastic impact on their economic viability. The independent tanker owners association, Intertanko, has calculated that at least 86 Greek flagged single hull tankers are at risk (of early withdrawal and scrapping) amounting to eight million tonnes of shipping.[13] The polemical nature of the debate between the

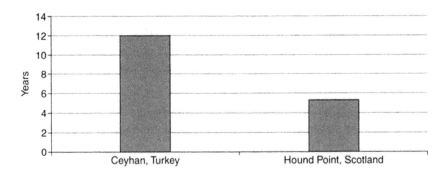

*Figure 19.1* Average tanker ages, Ceyhan and Hound Point (source: Lloyds Seasearchers).

Greek tanker owners and the state, advocates of a tougher policy on single hull European Union policy, is evident in Intertanko's statement:

> If one assumes not unreasonable average daily earnings of US$20,000 per day for each tanker, this amounts to the 'destroying' of more than US$1.3bn in earnings – definitely a very high price to be paying for one crack on the *Prestige*, more correctly, the failure of Spain to provide *Prestige* with a place of refuge.[14]

### 3 European maritime policy

For many years, the content of an extremely limited and developing Common Shipping Policy (CSP) concentrated on the attempts to reverse the phenomenon of flagging-out and progress the liberalisation of cabotage trade against a policy of flag restrictions. In the late 1980s, the debate included issues regarding the possibility of creating a European Registry of Shipping, 'Euros',[15] and an apparent concern (but lacked EU action) about safety issues.[16] In 1996, a 'maritime strategy document'[17] reviewed the outlook of maritime transport and focused on the future prospects of EU shipping, with a view to formulating the new guidelines of the CSP. At the same time, the strategy document declared the official abandonment of the earlier 1989 proposal for the creation of the Euros.

It was during the late 1990s that more attention was paid to the promotion of short sea shipping. The year 2000 was marked by a political agreement within the Council of Ministers, on the criteria regarding the inclusion of European ports in the EU plans on the Trans-European Transport Networks (TEN-T). The CSP was seen to be an integral part of the Common Transport Policy (CTP) and it could no longer be ignored.

In 1999, the European Commission identified the need to promote European shortsea shipping. This was justified on three counts:

1   the environmentally sustainable aspects of shipping;
2   undercapacity of the shortsea shipping sector, allowing for its growth and integration into the logistics supply chain; and
3   the capability of shipping to contribute to European economic integration.[18]

According to the Commission,[19] the achievement of environmental objectives through the CTP requires the introduction of regulations and economic measures promoting environmentally friendly transport modes without, at the same time, causing distortions in competition between the various modes. Major tanker accidents, with ensuing pollution disasters – *Torrey Canyon* (1967), *Amoco Cadiz* (1978) are evidence of the need for a strong European CSP that is high on the policymaking agenda. In terms of creating a competitive transport policy, it can be argued that the abolition of protectionist cabotage policies has been a successful area of European integration, which has impacted upon the maritime sector.

Paixâco and Marlow have identified the crucial role of European shipping policy in promoting a quality shortsea and coastal sector:

> The introduction of Port State Control (PSC) was a first step in that direction. The next will be the compliance with international regulation. If that is achieved maritime transport will get its dividends. It will change its image, and that, within a marketing context, is the basis for promoting the shifting of freight from road to sea.[20]

Much of the CSP within Europe has been geared towards the deepsea sector market, with attention given to the decline of the fleets of the traditional maritime nations and the rise of flags of convenience.[21] The decline of the European shipping industry, in itself, has an inverse relationship to the policymakers' attention to maritime safety. The shift in maritime operations to developing nation flags and crews has raised questions over quality and relatively high casualty rates for flags of convenience. In addition, contained in the shortlist for an extended Europe there are a number of nations whose vessels do not currently meet the necessary shipping standards of the EU (see next section). However, once these nations officially become members of the EU, there will be a large boost to the community-flagged fleet.

## 4 The extended community

Regulation on the passage of vessels along the European coastline has always been compatible with the Grotian principle of *Mare Liberum* – freedom of the seas. However, a number of nations on the accession list have questionable shipping safety standards. In 1985, Cyprus was signalled out as a sub-standard flag with high casualty rates and the violation of internationally accepted seafarer employment standards.[22] With regard to the effective implementation of the International Safety Management (ISM) Code, during the first five months of 2002, the Cyprus Authorities revoked the Documents of Compliance of three companies managing four Cyprus flag vessels in total.[23] One of these Documents of Compliance has been reinstated as a result of successful additional audits on shore and on board the vessels.

Cyprus has made considerable strides to improve safety in recent years. By the number of inspectors and through the global network of flag state Inspectors, standards have increased significantly during the last four years. They carried out 593 inspections in 2001, compared to 527 in 2000, 369 in 1999 and 166 in 1998.[24] However, the statistics still show that Cyprus has relatively high casualty figures, and is classed by the IMO as a middle risk flag. In order to complete preparations for membership, Cyprus' efforts now need to focus on, amongst other things, the completion of alignment and implementation of maritime transport legislation.

Cyprus is not the only country where further legislative work is still needed. For example, for Latvia to complete its membership preparations there is a need to concentrate on flag state inspection, classification societies and marine equipment. Whereas for Lithuania, further attention needs to be paid to strengthening the maritime safety authorities and ensuring that their independence can be guaranteed.

Other countries have successfully improved their shipping safety standards by introducing new legislation. Poland can be used as an example of this. In June 2002, the Maritime Code, the law on maritime safety and the amendment to the law on seaports have both entered the Polish statute books. Poland vessel detention rates, as a result of Port State Control, are below the average for EU flagged vessels (see Table 19.1 below).

Estonia adopted the Maritime Safety Act in December 2001. The Act has enabled the Estonian Government to implement important parts of the *acquis*, such as the level of maritime safety and the responsibilities of a shipowner. However, Estonia still needs to pay particular attention to improving the flag state performance of the Estonian fleet (see Table 19.1 below).

Malta is currently on the Paris Memorandum of Understanding's (MOU) black list. Between January and December 2001, 21 ships representing 270,257 gross tonnage were refused Maltese registry, while 40 ships, representing 468,933 gross tonnage, were removed from the ship registry due to technical non-conformity.[25] Substantial parts of the *acquis* still need to be transposed and further efforts need to be made to implement the *acquis*. For Malta to meet the objective of its removal from the black list, it needs to further improve its efforts concerning safety standards, notably through adequate staffing and training of the Malta Maritime Authority.

Even those countries that are landlocked, such as Hungary and Slovakia are expected to align themselves with the *acquis* on Port State Control. For Hungary,[26] Slovakia[27] and Slovenia[28] alignment with most of the

*Table 19.1* Port state control detention rate of new accession countries in 2001

| Country | % of detained vessels 2001 (EU average 3.1%) |
| --- | --- |
| Turkey | 24.5 |
| Romania | 23.5 |
| Bulgaria | 15.7 |
| Malta | 9.5 |
| Estonia | 7.9 |
| Lithuania | 7.4 |
| Latvia | 5.0 |
| Poland | 1.6 |

relevant *acquis* in this sector is almost complete. According to 2001 statistics under the MOU, Table 19.1 above shows the detention rate of vessels flying the individual nations' flag, compared to the EU average for EU flagged vessels of 3.1 per cent.

As the figures in the table show, those countries which will not be joining the EU in 2004 (Turkey, Bulgaria and Romania) still have a long way to go before their shipping fleets meet the necessary standards of the EU. For example, all three nations' flags are on the black list of the Secretariat of the MOU on port state control. The Bulgarians have experienced a delay in adoption of amendments to the Merchant Shipping Code, which has slowed further transposition of the *acquis*. Inspections of Bulgarian flag vessels as well as port state control inspections are at present carried out by 42 qualified inspectors.

Romania has re-organised its institutions in order to avoid duplication and excessive dilution of responsibilities – but whether these changes lead to an actual improvement in safety standards remains to be seen. Further significant legislative alignment can be noted for Romania. A framework law on Maritime and Inland Waterway Transport was adopted in June 2002, which will form the legal basis for alignment with a number of provisions of the *acquis* regarding both non-safety and maritime safety issues. The law also allows for the set up of a new body, the Romanian Naval Authority, through the merger of the existing bodies. It will be fully responsible for flag state implementation, flag state control and port state control. Furthermore, the authorities have continued their policy of 'cleaning up' their national shipping registry, and the total number of vessels flying the Romanian flag has decreased to 43. However, these positive developments have not yet resulted in acceptable detention rates.

Turkey has not lifted its restrictions applied to Cyprus-flagged vessels and vessels serving the Cyprus trade. Market access to coastal trade remains reserved solely for Turkish-flagged vessels. Turkey needs to enhance maritime safety and considerably intensify its efforts to decrease the detention rates for inspected ships.

## 5 The coastal and shortsea trades

The promise of coastal and shortsea shipping in the extended European community is considerable. The importance of some degree of public policy was argued for at the outset of this chapter. Having outlined the obstacles to be overcome in gaining freight transfer in the UK, the future implementation of European transport policy will favour water transport. the two critical areas are:

1   the carbon tax imposition; and
2   the identification of the externally generated environmental costs of transport.

Both policy areas are likely to serve to the detriment of road haulage economics. The White Paper reports that 28 per cent of $CO_2$ emissions are now transport related and transport energy consumption is increasing. If nothing is done between now and 2010, there will have been a 50 per cent rise in $CO_2$ emissions between 1990–2010, whereby 84 per cent of transport emissions are related to road transport.[29] Given the sharp contrast in energy efficiency per tonne – kilometre between road and water, the imposition of such 'green' taxes can only serve to benefit the latter transport mode. The end result of such policy helps reduce the critical distance/volume thresholds where water transport becomes competitive with road haulage.

The twin problems of road congestion and vehicle fuel emissions is seen as a threat to the concept of an economically integrated and environmentally discerning European Community. Worsening journey transit times and the economic inefficiency of lorries delayed by congestion is now being challenged. For the first time in the short history of the motor lorry its supremacy is beginning to be questioned in the world of European politics. Partly this can be seen as the result of the critical levels of rapidly growing road transport. To a large extent the transport market has failed to ensure a workable solution to the transport problem. The preoccupation with cars has been at the expense of public transport, buses and passenger trains; the emphasis on lorries has proved detrimental to inland and coastal shipping and railfreight. The desire of European policymakers is recognisable as an attempt to divert some traffic away from roads. The EU has recognised that three modes have the spare capacity to achieve a significant shift away from road haulage:

1   railways;
2   inland shipping; and
3   coastal and shortsea shipping.[30]

In order to facilitate this shift in 1995 the Commission adopted three areas of action:

1   to improve the quality and efficiency of shortsea transport;
2   to improve ports and infrastructures; and
3   the inclusion of shortsea shipping within the CTP external relations framework.[31]

The 2001 European Conference of Ministers of Transport report, 'Short Sea Shipping in Europe', has identified the importance of shipping within the context of European cohesion in the way it:

•   promotes European trade competitiveness;
•   maintains vital transport links;
•   decreases unit costs of transport;

- facilitates Eastern European integration; and
- relieves congestion from land-based networks.[32]

The problem, however, is how can quality shipping compete successfully against:

- a dynamic road haulage industry; and
- low cost/low quality tonnage.

The gridlock, global warming, associated with road haulage has led to policymakers heralding the shipping alternative. However, the economic realities of trade are that shipping costs are not always competitive with road haulage. The economies of scale derived from the larger tonnage movements by sea can easily be eroded by handling and transhipment costs. The road haulage industry has the advantages of flexibility and an aggressive entrepreneurial stance. The shipping option is only now beginning to be taken seriously as an option to limit the excesses of the road solution. The 'green highway', therefore, offers a more environmentally sustainable, less crowded route around the British Isles and Ireland. Although there has been a significant growth in European shortsea shipping tonne-kilometre activity this has been outstripped by road haulage: between 1990 and 1999 shortsea shipping grew by 30 per cent, an annual average of 2.9 per cent; in the same period road haulage grew by 41 per cent, an annual average of 3.9 per cent.[33] Italian maritime economists, Musso and Marchese have identified the nature of shortsea growth with focus on developing captive markets, rather than new markets, which would bring a true shift from land transport to water.[34] In the UK and Ireland, however, the 'captive' trades have also been prone to road haulage competition.

The competitive situation is likely to increase, as low cost nation suppliers of road haulage services enter into the European mainstream market. The breakdown of the Soviet trading bloc cost and the privatisation of road haulage fleets has led to a 'Wild East' situation. The early days of the post-Soviet era were characterised by low running costs and profitable returns on loads. One industry watcher has commented that in this period it was possible to drive from the Slovak/Ukranian border to Alma Ata in Kazakhstan with only a case of beer to cover all costs.[35] The rewards for such haulage runs quickly induced over supply in the market; from 1995 onwards over-capacity of truck and drivers has led to falling rates and the ITF has drawn attention to the large amount of East European drivers financially stranded in Western Europe. Unable to meet fuel bills, the drivers have been ordered to seek return cargoes or stay put.

The attraction of employing East European drivers at rates as low as £15 per day has not been ignored by West European hauliers.[36] Under such conditions it is inevitable that rates will fall as haulage firms and their drivers seek out return loads at competitive prices.

This raises questions on how shipping can compete in the open market. Already there have been concerns over the 'lean manning' in the feeder trades. The groundings of two feeder ships, the *Cita* and the *Coastal Bay* in British waters has been linked to excessive fatigue.

As well as the incessant pressure from road haulage, the quality operators in the European trades face the encroachment of low cost/low quality operators in the market. Such actions have led to the displacement of large amounts of ageing from the domestic trades of the former Soviet Union. In particular, the river class ships of the Volga have been forced into the North Sea and Mediterranean waters.

## 6 The economic pressures from these dual forces

In the dry bulk trades, in particular, the impact of competition has been felt. One of the brighter sectors of UK shipping, post-1980, was that of coastal and shortsea. Traditional British companies from coastal regions – Everards of Greenthithe, Kent, Lapthorn's of Rochester, Kent, Union Transport of Bromley, Kent, Fishers of Barrow in Furness, Cumbria and Crescent Shipping of Rochester (now Southampton). These companies had been able to retain – in some cases increase – their market share whilst still remaining under the British flag. In particular, these companies have been successful in the highly competitive quality-oriented clean petroleum products (CPP) market. Crescent Tankships, Everards and Fishers have attained sizeable long-term contracts with oil majors for the dispatch of oil products ex refinery.

Demonstrating the difficulties of trading in the low value bulk sector was the 2002 announcements of both Crescent Shipping and Fishers of their withdrawal from the dry bulk sector. Both companies retained tanker vessels dedicated solely to the CPP trades. Also in 2002 came the announcement that the major operator, F.T. Everard, was withdrawing from the dry bulk market.[37] Everards had consistently reduced their exposure to the risky dry bulk market in the previous decade and focused on the CPP sector. The trend towards withdrawal was further endorsed by the January 2003 announcement that the hitherto successful Lapthorn Shipping was placing its whole fleet of 19 general cargo ships on the sale and purchase market.[38] Between 1990 and 2003, the following UK dry bulk fleets have been sold:[39]

* Crescent Shipping    17 vessels;
* F.T. Everard    17 vessels;
* James Fisher    6 vessels;
* Lapthorn Shipping    19 vessels.

At a time when European transport policy is calling for increased utilisation, of coastal and shortsea shipping, the expertise leading traditional owners is lost to the market. The combined competitive pressures of the open market in shortsea and coastal shipping and intense road haulage competition has made shipowning untenable for UK owners in this sector.

## 7 The Baltic trades

The vast opportunities for the tanker industry arise from the Baltic oil fields. As an example of the growth potential is the current $3.7 billion project to develop oil terminals in Russia, Gulf of Finland port, Primorsk. The project is intended to both increase tanker sizes from 100,000 to 150,000dwt and will boost oil flows by 45m. tonnes.[40] The problem, however, is how to link the Russian, Latvian and Lithuanian oilfields with the main world markets. The main trading routes for tankers above 20,000dwt are via the complex of Danish Islands. Additionally, oil flows from the Northern fields have to deal with the adverse winter conditions, excessive icing problems. Additionally, the route takes loaded tankers close to the British, Irish and continental mainland coastlines.

The economic versus environmental dichotomy manifests in a number of ways. Concerns over tanker groundings and the potential for collisions have prompted calls for compulsory pilotage in the Kattegat-Baltic region. Such calls have been opposed by Russian oil exporters as an unnecessary economic burden.

The question of ice hazards in the Northern Fields raises the question of the suitability of tankers. The Finns have concerns over the loadings of two Suezmax tankers, *Stemnista* and *Minerva Nounnou*. It has been argued that these double hull vessels, built in 2000, do not comply with the ice specifications of the Gulf of Finland. In February 2003, it was reported that the vessels' charterer was reluctant to release these vessels from their charter party obligations.[41] The dichotomy between costs and quality is apparent in this area of concern. The ice strengthened tanker costs up to 33.3 per cent of the standard newbuild. Comparative capital costs for a new standard Aframax and an A1 ice strengthened Aframax tanker in March 2003 were quoted as:

- Standard tanker:          $36.5m
- A1 ice strengthened:      $49.5m[42]

Finnish concerns are that tankers not specifically designed for the Northern Winter are vulnerable in extreme ice conditions. *Fairplay* has reported on the differing approaches to safety between Finland and Russia towards safety issues with Finland favouring high cost, environmentally friendly, tankers designed specifically for the ice borne routes, and Russia having a less cautious approach:

The Finns tend to play by the book: there cannot be many other countries in the world where people stop at red lights in an empty crossing until the lights turn green again. They believe that their neighbours in the East are prone to cut corners with little care for the consequences.[43]

In addition to the concerns over ice strengthening, there has been consternation in Denmark over the increasing volumes of tankers transiting the difficult channels of the Kattegat. A number of tanker groundings have led to calls for compulsory pilotage; however this has been opposed by Russian oil interests on the grounds of increased costs.[44]

The aftermath of the 2002 *Prestige* disaster – following so closely after the *Erika* sinking – has focused attention on chartering practice. There were revelations that certain charterers concentrated on older tonnage which facilitated lower freight rates thus higher profit margins for the traders. It has been estimated that savings of up to $5,000 per day were possible by chartering such elderly vessels as the *Prestige*. The company concerned has particularly focused on older tonnage, with the average age of tankers increasing from 11 to 19 years in the last three years, very much in contrast to the general trend in the tanker market.[45]

## 8  The Caspian/Black Sea trades

The exploitation of the landlocked Caspian field presents the oil majors with a major logistics challenges. Additionally, the oil region and its pipeline routes to the Black Sea and Mediterranean are via areas of political instability. Loadings in the Black Sea ports – Novorossiysk, Tuapse, Supsa, Batumi are the preferred choice for the Russian Government. US oil interests have supported the 1,038 mile pipeline from Baku, on the shores of the Caspian Sea, to the Turkish port, Ceyhan, in the Mediterranean. This will boost oil flows by one million barrels per day.[46] By 2015, projected oil flows on routes from the Caspian are set to grow by around 4.5 million barrels per day.[47]

In addition to the political uncertainties of the Caspian region, the Black Sea route is via the shipping bottleneck, the Bosporus and Dardanelles Straits. The restrictions of the narrow passage have raised questions over the security and safety of the route. Events in the storms of February 2003, culminating with a tanker grounding in both Straits led to a critical build up of oil supplies in the Russian Black Sea port, Novorossiysk. The prospects of increased tanker tonnage movements in the restricted seaways and across the busy shipping lanes of the Mediterranean must raise concerns over safety. This is particularly the case given the enduring problem of ageing tankers in the region.

## 9 Conclusion

In 2004, the Commission will present a more extensive review of the Common Transport Policy, one of the issues it plans to cover in the review is how best to link the outlying regions on the European continent more effectively and connecting the networks of the candidate countries to the networks of EU countries.

In spite of maritime disasters, the potential of shipping as a substitute for more environmentally damaging or costly transport modes is clear. However, the risks and possible consequences of maritime accidents need to be minimised. It needs the involvement of all actors from all levels of decision making to accept the challenge so that effective combinations of strategies can be adopted in particular situations to address the key problems identified.

This chapter has sought to identify some of the more problematic areas of combining shipping growth with maintaining and, moreover, improving safety standards. The intention has been to inform the debate on maritime standards of some of the future challenges faced by policymakers. The considerable opportunities for competitive and environmentally sustainable coastal and shortsea shipping within the Union can be attained. However, there needs to be recognition of the economic forces at work which could negate the environmental benefits of a switch from roads to water, if maritime standards are diluted by the need to reduce costs.

The exploitation of the considerable oil supplies of the Caspian and Baltic provides attractive opportunities for growth in the former Soviet Union republics. Efficient and competitive tanker shipping is an important pre-requisite of developing this trade. And it is evident that there is a division of economic and environmental opinion between European coastline nations and the oil exporters. It has been made evident that, following the intense level of concern generated by such tanker disasters as the *Erika* and *Prestige*, European policymakers will face major challenges in attempting to balance the economic-environmental dichotomy, the need to weigh the future wellbeing of both these trades and the marine environments they place at risk.

### References and notes

1 CEC, 2001, European Transport Policy for 2020: Time to Decide, (COM(2001)370), Office for Official Publications of the European Communities, Luxembourg.
2 10 of the 13 applicant countries have been invited to join the EU by May 2004. Cyprus, Czech Republic, Estonia, Hungary, Latvia, Lithuania, Malta, Poland, Slovakia and Slovenia will join in the first round. The EU executive has set a proposed target date of 2007 for Bulgaria and Romania to join. Turkey has been told that it must wait at least two more years before starting talks on joining the group. See http://www.news.bbc.co.uk, 17/02/03.

3 CEC, 1997, Green Paper on Seaports and Maritime Infrastructure (COM(97)678).

4 CEC, 1997, Proposal to Integrate Ports and Terminals into the TEN-T Guidelines, (COM(97)681).

5 CEC, 1998, Report on the Implementation of the Guidelines and Priorities for the Future – Decision 1692/96/EC.

6 Transport: the Commission formulates the political guidelines of the White Paper on the future of the Common Transport Policy, IP/01/1008, Brussels, 18.07.01.

7 CEC, 2001, European Transport Policy for 2020, Op. Cit.

8 IP/01/1008.

9 www.eia.doe.gov Energy Information Administration, Caspian Sea Region: Oil Export Options, July 2002.

10 Aspinwall, M.D., 1995, *Moveable Feast: Pressure Group Conflict and the European Community Shipping Policy* (Aldershot: Avebury).

11 Brodin, A., 2001, Going West! Russian reasons to focus on ports in the Gulf of Finland, a paper presented to the Eighth World Conference on Transport Research (WCTR), Seoul, July 2001.

12 Donaldson, Safer ships. cleaner seas – a reflection on progress, The Wakeford Lecture, University of Southampton, 26.2.96.

13 Single hull could cost owners billions, *Lloyd's List*, 30.1.03.

14 *Loc.cit.*

15 CEC, 1989, A Future for the Community shipping industry: measures to improve the operating conditions of Community shipping (COM(89)266), final 3 August.

16 Chlomoudis, C. and Pallis, A., 2002, European Union port policy: the movement towards a long-term strategy (Cheltenham: Edward Elgar).

17 CEC, 1996, Shaping Europe's maritime future – a contribution to the competitiveness of Europe's maritime future, (COM(96)84), final 13 March.

18 CEC, 1997, Communication from the Commission to the European Parliament, the Council, the Economic and Social Committee and the Committee of the Regions, *The Development of Short Sea Shipping in Europe: A Dynamic Alternative in a Sustainable Transport Chain. A second two yearly progress report*, COM (9) 317 (Luxembourg: Office for Official Publications of the European Communities).

19 CEC, 1992, Communication and legislative proposals concerning the creation of a European combined transport network, (COM(92)230), final, 11 June 1992.

20 Paixâco, A.C. and Marlow, P.B., 2001, A review of the European Union shipping policy, *Maritime Policy & Management*, 28, 2, 198.

21 Aspinwall, M.D., 1995, Op. Cit. 5.

22 Metaxas, B.N., 1985, *Flags of Convenience* (Aldershot, Hampshire: Gower Press) 25.

23 Regular Report, 9 October 2002, http://europa.eu.int/comm/enlargement/cyprus/index.htm.

24 Ibid.

25 Regular Report, 9 October 2002, http://europa.eu.int/comm/enlargement/malta/index.htm.

26 Regular Report, 9 October 2002, http://europa.eu.int/comm/enlargement/hungary/index.htm.

27 Regular Report, 9 October 2002, http://europa.eu.int/comm/enlargement/slovakia/index.htm.

28 Regular Report, 9 October 2002, http://europa.eu.int/comm/enlargement/slovenia/index.htm.
29 IP/01/1008.
30 CEC, 1993. The future development of the common transport policy. A global approach to the construction of a community framework for sustainable mobility, *Bulletin of the European Communities, Supplement 3/93* (Luxembourg: OOPEC) 24.
31 Ross, F.L., 1998, *Linking Europe: Transport Policies and Politics in the European Union* (Westport: Praeger) 168.
32 Papadimitriou, S., 2001, European Conference of Ministers of Transport Report: *Short Sea Shipping in Europe*, Paris ECMT, 10.
33 Musso, E. and Marchese, U., 2002, Economics of shortsea shipping, T.H. Grammenos (ed.) *The Handbook of Maritime Economics and Business* (London: LLP) 285.
34 Ibid.
35 Will trucking companies survive the year 2000? www.geocities.com/WallStreet/Floor/5990/art3.html.
36 Socialist truck drivers press release, 6.6.02. www.socialisttruckdrivers.org.uk.
37 Grey, M., 2002, Union transport and FT Everard link dry bulk fleets, *Lloyd's List*, 13.3.02, 1.
38 Speculation mounts over Lapthorn sell-off, *Lloyd's List*, 27.1.03.
39 Derived from *Lloyds Register of Shipping List of Shipowners*.
40 Brodin, A., 2001, Going West! Russian reasons to focus on ports in the Gulf of Finland, 16.
41 Tanker *Steminista's* sister ship approaching Primorsk oil terminal, *Heingin Sanomat*, 5.2.03.
42 Source: George Moundreas & Company. S.A. Shipbrokers, Piraeus.
43 Scheurer, B., Tanker grounding in Kattegat prompts new Baltic pilots plea, *Lloyd's List*, 3.3.03.
44 An icy issue becomes hot news, *Fairplay International Shipping Weekly*, 20.2.03.
45 Crown cashes in with older tankers, *Lloyd's List*, 27.11.02.
46 Caspian Sea region: oil export options, July 2002, www.eia.doe.gov.
47 Caspian Sea region: oil export options, July 2002, 8.

# 20 Maritime legislation
## New areas for safety of life at sea

*K.X. Li and J. Wonham*

## 1 Introduction

The *Titanic* tragedy in 1912 claimed 1,489 lives, and made the world realize that, due to the nature of shipping, maritime safety can not be achieved without international co-action and co-operation.[1] This resulted in the first International Conference on Safety at Sea in 1913–14, and the first international convention on maritime safety, the Convention for the Safety of Life at Sea (SOLAS), signed on 20 January 1914. Afterwards, and especially in the last two decades, safety regulations have been adopted intensively within the International Maritime Organization (the IMO) and the International Labour Organization (the ILO) regimes. However, implementation of these regulations is still inadequate. The open-registry scheme has been recognized as a weak link in the maritime safety chain.[2] There emerges an obvious duty for governments to take care of the safety, health and welfare of seafarers. The study reveals that heavy losses of life occurred at sea as a result of personal injuries, crimes at sea and coastal ferry disasters, and which areas need to be addressed by improvement of maritime legislation and proper enforcement of existing regulations.

## 2 Data

Worldwide data on merchant vessels, e.g. numbers of ships, sizes, types, numbers of new building and accidental total losses, have been available for many years. Complete data on the employment of seamen is still not available on a worldwide scale, and not even at national levels in some maritime nations. In 1997, the number of seamen in the UK, a traditional maritime nation, was based on estimation.[3] Calculations of maritime labour supply and demand are based on questionnaires.[4] The fatalities and injuries to seamen worldwide have to be approximated, based on the data of individual fleets.[5] Without 'safer, healthier and happier seamen', the aim of 'safer ships and cleaner oceans' can not be fully achieved.

Kitchen[6] correctly concludes that 'of all sections of the community, seafaring men ... have been the most ignored and therefore the worst treated'. One good piece of evidence of this is the non-availability of necessary seamen related data, e.g. their total number, accidents and fatalities, which are essential data for research. 'This lack of academic attention is a shame in that the growth of international economic interdependence has increased the importance of the maritime industry world-wide',[7] which is especially true in the case of seamen working on board 'open registry'[8] ships, since 'the open-registry fleets were operating without effective government control or regulation' either by flag states or beneficial states.[9] This results in not only the poor safety record of open-registry fleets, but also the non-availability of data relating to working conditions, in general, and injury and death to seamen on open-registry ships.[10]

In this study, data of fatalities to UK seamen has been collected from publications of the UK Government, i.e. the Department of Trade (1962–88), and later the Marine Accident Investigation Branch (MAIB) under the Department of Transport (1989–97). MAIB is an independent unit within the Department of Transport, established under the Merchant Shipping Act to undertake maritime accident investigation and to publish the results of the investigations. The data covers seamen working on UK registered vessels, but does not include UK seamen on non-UK registered vessels. Data on accidental mortality to worldwide seamen has been collected from the Institute of London Underwriters (ILU) (1986–95).

For a calculation of fatal accident rates and mortality rates, the population of seamen at risk should be established. The number of seamen is not as simple a concept as it appears. The number of seamen may sometimes refer to 'employed seamen', which are the number employed by shipping companies, including seamen both on board ship and on shore for holidays or illness. UK data shows that the average ratio between seamen posts and employed seamen is 1:1.2.[11] It can also be confused with 'active seamen', and 'qualified seamen'. Maritime authorities usually have records of these qualified (registered) seamen, but usually have no records on 'active seamen', 'employed seamen', or 'population at risk'. It is worth noting that most previous studies reviewed took qualified (registered) seamen, which is easier to obtain from maritime authorities, as the population at risk, which would be greater than the number of employed seamen. The latter, in turn, would be larger than the number at risk. Obviously, those fatality mortality rates calculated based on 'qualified seamen' as 'the population at risk' would underestimate the reality.

In order to reach a more accurate conclusion on seamen mortality rates and fatal accident rates at work, this study uses the number of posts as the number of seamen at risk, which is calculated in accordance with Li and Wonham.[11] In this study, fatalities refer to deaths as a result of personal accident and ship causalities. Mortality refers to deaths from all causes, including those from diseases. Fatality (mortality) rates can be calculated in equation (1):

$$\text{Rate} = 1000 * \text{annual total fatality (mortality)/posts (‰)} \qquad (1)$$

## 3 Analysis of UK data (1962–88)

The study collects and examines 5,389 mortality cases to UK seamen from 1962–88, i.e. ~200 cases per year (Table 20.1). The mean fatality rate is 1.28‰, and mean mortality rate is 2.81‰ per year. All mortality cases can be divided into four groups according to their causes.

The first type is caused by 'vessel casualties', which includes founderings/floodings, strandings/groundings, collisions/contacts, capsizing/listing, fires/explosions, machinery damage, heavy weather, missing vessels, and other casualties to ships. Of 5,389 recognized mortality cases, 572 cases (10.61 per cent) were caused by ship casualties, 21 cases per year on average.

The second type is 'personal accidents', e.g. injuries sustained as a result of slips and falls on and over board, during the operation of hatches and winches and other machinery, using ropes and hawsers, exposure to noxious substances, and those reported as missing at sea. There are a total of 1,749 cases falling into this category, at a mean of 65 per year, and forming 32.5 per cent of all mortality cases (categories 1–4), and 63.7 per cent of all fatal accidents (categories 1–2) (see Table 20.1).

From categories 1 and 2, one can derive a fatal accident number, which varies greatly from year-to-year, the lowest being 0.23‰ and the highest 3.48‰, due to the fluctuations in the number of deaths arising from vessel casualties. The mean annual rate is 1.28‰. Compared with other UK high risk industries (1959–64),[12] the mean of fatal accident rates to seamen is higher than that of fishing (1.00‰), 3 times that of coal-mining (0.44‰), 5.5 times that of the construction industry (0.23‰), and 25 times that of the manufacturing industry (0.05‰). This gives some indication of the hazards of seafaring.[13]

The third category of cause is as a result of 'suicide and homicide' incidents. A total of 348 suicide cases were identified: i.e. an average of 13 cases per year, giving a mean annual suicide rate of 0.16 per cent. This may suggest the stressful nature of conditions at sea. A person distressed to the extent of committing suicide certainly would not be able to perform his duty or function properly, but could also even become a hazard to safety. However, it should be noted that the numbers of both suicides and homicides have consistently fallen over the years, which may be attributable to both the improvement in working and living conditions on board ships and the decline in numbers of UK seamen. The fourth category is mortality from 'diseases': 2,640 mortality cases, or 49 per cent of the total, were identified as the result of diseases, an average of 98 per year. There is no information in the data obtained as to the details of illnesses and diseases leading to the demise of seamen. Studies,[14] however, showed that Chronic Heart Disease (CHD) was the principal natural cause of death amongst seamen, and suggested the proportion of deaths from this cause is higher than in other occupations because of extensive stress and fatigue on board ships.

Table 20.1 Mortality of UK seamen (1962–88)

| Causes | Vessel casualty[a] Category 1 | Personal accident[a] Category 2 | Suicide[a] Category 3 | Homicide[a] | Disease[a] Category 4 | Total | Posts[b] | Fatality rate (‰/year) | Mortality rate (‰/year) |
|---|---|---|---|---|---|---|---|---|---|
| 1962 | 34 | 130 | 20 | 8 | 225 | 417 | 103,614 | 1.58 | 4.02 |
| 1963 | 8 | 103 | 32 | 5 | 203 | 351 | 97,669 | 1.14 | 3.59 |
| 1964 | 28 | 115 | 22 | 5 | 171 | 341 | 92,597 | 1.54 | 3.68 |
| 1965 | 17 | 101 | 22 | 3 | 173 | 316 | 89,538 | 1.32 | 3.53 |
| 1966 | 68 | 103 | 18 | 5 | 141 | 335 | 86,017 | 1.99 | 3.89 |
| 1967 | 45 | 91 | 27 | 9 | 153 | 325 | 82,546 | 1.65 | 3.94 |
| 1968 | 9 | 66 | 19 | 2 | 136 | 232 | 79,364 | 0.95 | 2.92 |
| 1969 | 13 | 79 | 19 | 1 | 162 | 274 | 77,536 | 1.19 | 3.53 |
| 1970 | 32 | 72 | 30 | 5 | 137 | 276 | 77,621 | 1.34 | 3.56 |
| 1971 | 14 | 63 | 31 | 5 | 134 | 247 | 77,270 | 1.00 | 3.20 |
| 1972 | 75 | 59 | 25 | 5 | 127 | 291 | 76,100 | 1.76 | 3.82 |
| 1973 | 19 | 83 | 13 | 2 | 110 | 227 | 75,387 | 1.35 | 3.01 |
| 1974 | 13 | 75 | 17 | 7 | 118 | 230 | 75,229 | 1.17 | 3.06 |
| 1975 | 32 | 71 | 22 | 4 | 105 | 234 | 73,400 | 1.40 | 3.19 |
| 1976 | 13 | 58 | 4 | 6 | 65 | 146 | 70,900 | 1.00 | 2.06 |
| 1977 | 3 | 52 | 6 | 0 | 71 | 132 | 69,600 | 0.79 | 1.90 |
| 1978 | 8 | 84 | 4 | 2 | 74 | 172 | 66,000 | 1.39 | 2.61 |
| 1979 | 16 | 55 | 5 | 0 | 55 | 131 | 60,400 | 1.18 | 2.17 |
| 1980 | 45 | 79 | 4 | 1 | 67 | 196 | 57,900 | 2.14 | 3.39 |
| 1981 | 12 | 44 | 3 | 1 | 51 | 111 | 53,700 | 1.04 | 2.07 |
| 1982 | 18 | 49 | 2 | 1 | 39 | 109 | 47,500 | 1.41 | 2.29 |
| 1983 | 1 | 19 | 1 | 0 | 41 | 62 | 40,000 | 0.50 | 1.55 |
| 1984 | 3 | 16 | 2 | 0 | 19 | 40 | 34,500 | 0.55 | 1.16 |
| 1985 | 1 | 7 | 0 | 0 | 23 | 31 | 34,500 | 0.23 | 0.90 |
| 1986 | 0 | 15 | 0 | 0 | 16 | 32 | 24,800 | 0.60 | 1.29 |
| 1987 | 39 | 47 | 0 | 1 | 10 | 97 | 24,703 | 3.48 | 3.93 |
| 1988 | 6 | 13 | 0 | 1 | 14 | 34 | 22,629 | 0.84 | 1.50 |
| Total | 572 | 1,749 | 348 | 80 | 2,640 | 5,389 | 1,771,020 | 34.54 | 75.76 |
| Mean | 21 | 65 | 13 | 3 | 98 | 200 | 65,593 | 1.28 | 2.81 |
| % | 10.61 | 32.46 | 6.46 | 1.48 | 48.99 | 100 | | | |

Note
a UK Department of Trade[27]   b 1975–1986[28]; 1987–1990[29]; 1962–1974 and 1985–1986.[11]

As to the causes of the 3,559 accidents (both fatal and non-fatal) from 1989–97, the study finds that slips and falls (46 per cent), manual handling (20 per cent), and machine operations (19 per cent) are the main causes, forming 85 per cent of the total accidents identified.

## 4 Analysis of worldwide data (1986–95)

ILU, apparently, is the only source that keeps records of fatalities as the result of ship casualties (category 1) from 1986–95. The annual average fatalities for this category is 688. However, there is no data available on deaths from personal accidents (category 2), from suicides and homicides (category 3), or from diseases (category 4). Based on ILU figures on mortality from ship casualties (category 1) and the mortality pattern of UK seamen, i.e. category 1 constitutes 20.6 per cent of all fatalities (categories 1–3), and 10.5 per cent of all mortality cases (categories 1–4).

Accordingly, one can arrive at the world annual fatalities = 688/20.6 per cent = 3,340/year, and mortality cases = 688/10.5 = 6,552/year. As the world average posts is 1,003,592,[11] accordingly, the average annual fatality mortality rate for world seamen is 3.33‰, and the average annual mortality rate is 6.53‰, which are about two times higher than those of UK seamen (1.28‰ and 2.81‰, respectively, see Table 20.1).

## 5 Estimations of fatalities to open-registry seamen

Due to non-availability of data on fatalities and injuries to open-registry seafarers, estimates of such numbers would be useful for both practical and academic purposes. This can be done by the improvement of Goss *et al.*'s method.[5] First, Goss *et al.*'s study collects only ten-year data (1979–88). This study collects 35-year data (1962–96). Second, the formula in their study can only describe a perfect situation. According to this formula, if there are no total losses in certain years, there will be no fatalities or injuries in that fleet, which is not generally the case. To compensate this, a constant is calculated in this study. Third, Goss *et al.*'s study only uses ship total losses in g.t. as the safety indicator. In this study, ship total loses in number is also considered in order to have a more accurate result.

Outputs of Linear Regression in SPSS/PC show that both ship total losses in number and in g.t. are positively correlated with seamen's fatalities. Fatalities with ships total losses number, gives the multiple $R_1$ as 0.73 (adjusted $R_1 = 0.51$, significant $< 0.001$); whilst for ship total losses in g.t., gives the multiple $R_2$ as 0.52 (adjusted $R^2_2 = 0.25$, significant $< 0.01$). This suggested that ship total loss number as an indicator can be more reliable than total losses in g.t. Accordingly:

$$\text{Fatality}_1 = C_1 + B_1 * \text{Loss No.} \qquad (2)$$

where Constant $C_1 = -8.028\,396$, coefficient $B_1 = 7.199\,404$, and

$$\text{Fatality}_2 = C_2 + B_2 * \text{Loss g.t.} \tag{3}$$

where $C_2 = 38.591\,214$, $B_2 = 0.001\,185$.

Accordingly, with different weighting according to their different coefficients, an adjusted fatality figure can be produced from $\text{Fatality}_1$ and $\text{Fatality}_2$ by equation (4):

$$\text{Fatality} = (R_1 * \text{Fatality}_1)/(R_1 + R_2) + (R_2 * \text{Fatality}_2)/(R_1 + R_2) \tag{4}$$

Based on the seamen posts on open registry ships, 217,654,[11] an annual fatality rate of 3.82‰ can be derived, which is higher than the world average of 3.33‰. This may imply open registry ships are more likely to cause damage to seamen and support the conclusion that the open registry scheme is a weak link in the maritime safety chain.[10]

## 6  Improvements of IMO legislation

After perusal of maritime legislation from the past four decades, some weakness can be observed from the IMO decision-making process. First, IMO regulations are usually made as reactions to certain events rather than a result of systematic examinations and analyses. Second, political consideration of what is seen as important to please public opinion has frequently dictated the IMO's priorities. Third, it becomes obvious that a disaster with multiple fatalities, which, however, occurs once in a long period of time, has always taken priority over numerous incidents involving one or two fatalities, which may add up to a greater loss of life over the same length of time.

### 6.1  Prevention of seamen personal accidents on board ships

Data shows that personal accidents form ~90 per cent of all mortality cases, much greater than that from vessel casualties. Therefore, prevention of personal accidents on board ship should be given a higher priority and properly dealt with by IMO conventions, which have a higher implementation rate than ILO conventions. However, it is noted that the IMO is generally not concerned with personal injuries on board ship. One of the aims and functions of the IMO is to promote the 'safety of life at sea', which, by definition, should not be limited to safety of ships and the environment. It could be argued that provisions for preventing occupational accidents should logically be put on to the IMO agenda.

A joint working group lead by Mr J.M. Shindler, the French Permanent Representative to the IMO, has urged that 'the problems of abandonment, personal injury and death of seafarers need urgent remedial action'.[15]

Hopefully, their suggestions can be heard and act as a springboard for change in the IMO working direction.[16]

### 6.2 Crimes at sea and abandoned seamen

There is nothing new about the abandonment of seafarers or the bankruptcy of shipowners. In the past four years (1996–99) alone 3,500 seafarers, relating to ~210 ship cases, have been left in strange lands without payments and any assistance whatsoever from bankrupted single-ship companies.[17] Some bankruptcies are genuine, whilst some owners just use single-ship-company and the open registry regime as a means to escape their obligations and liabilities toward seafarers or other creditors by disguising true identity and ownership.

Equally alarming are the statistics about crime at sea. Based on the data provided by the International Maritime Bureau (IMB),[18] it is noted in Table 20.2, which sets out all crime types from 1991–99 that there were a total of 2,452 crimes against seafarers at sea. During this period, 187 seamen were killed, 137 injured, 1,652 taken hostage, 343 threatened, and 133 assaulted. Safety at sea obviously can not be achieved without provision of safe and peaceful working conditions for seamen.

### 6.3 Bulk carrier disaster and nationality of deceased seamen (1990–98)

In the period from 1990–98, 37 bulk carriers, total 1,204,327 g.t., have constituted total losses: 23 ships (62 per cent) were in open registry ships, i.e. eight in Cyprus, seven in Malta, five in Panama, and three in Liberia.[19]

A total of 812 seamen died in these casualties. By examining data obtained from the International Transport Worker Federation (ITF),[20] Lloyd's Registry, the nationalities of 364 of the 812 deceased seamen have been identified. Of these seamen, 112 were Filipinos (30.8 per cent of the total), 81 Chinese (22.3 per cent), 51 Greek (14.0 per cent), 29 Indian (8.0 per cent), 24 Thai (6.6 per cent), 23 Poles (6.3 per cent), 13 Moldavian (3.6 per

*Table 20.2* Injuries and fatalities of seamen by crimes at sea (1991–99)

| Type of crime | 1991 | 1992 | 1993 | 1994 | 1995 | 1996 | 1997 | 1998 | 1999 | Total | Annual |
|---|---|---|---|---|---|---|---|---|---|---|---|
| killed | – | 3 | – | – | 26 | 26 | 51 | 78 | 3 | 187 | 21 |
| injured | 4 | 16 | 3 | 10 | 3 | 9 | 31 | 37 | 24 | 137 | 15 |
| hostage | 33 | 18 | 6 | 11 | 320 | 193 | 419 | 244 | 408 | 1,652 | 184 |
| threatened | 3 | 9 | 1 | 8 | 59 | 56 | 119 | 68 | 20 | 343 | 38 |
| assaulted | 2 | 12 | 4 | – | 2 | 9 | 23 | 58 | 23 | 133 | 15 |
| Total | 42 | 58 | 14 | 29 | 410 | 293 | 643 | 485 | 478 | 2,452 | – |

Source: IMB[30]

cent), nine Romanian (2.5 per cent), seven German (1.9 per cent), five Egyptian (1.4 per cent), three Yugoslav (0.8 per cent), two Moroccan (0.5 per cent), one Burmese (0.3 per cent), one Russian (0.3 per cent), one Turkish (0.3 per cent), one Ukrainian (0.3 per cent), and one Sri Lankan (0.3 per cent) (Table 20.3). This distribution of nationalities is in accordance with the main seamen supply countries set out in Li and Wonham.[21]

The heavy losses of bulk carriers in the late 1980s and early 1990s raised alarm bells at the IMO. As a result, in 1991 the Assembly of the IMO adopted an interim resolution,[22] concentrating on paying attention to the structural integrity and seaworthiness of ships and ensuring that the loading and carrying of cargo would not cause undue stress.[15] A new chapter (XII), 'additional Safety Measures for Bulk Carriers', was added to SOLAS ten years later, after the problem arose in the early 1990s.[22] The problem might have been delayed even further had it not been speeded up by a new report issued by the UK Government on the sinking of the UK flag bulk carrier *Derbyshire*.

*Table 20.3* Nationality of deceased seamen on bulk-carriers (1990–98)

| Nationality | Number | % | Developed nations (%) | Developed nations (%) |
|---|---|---|---|---|
| Philippine | 112 | 30.8 | – | 30.8 |
| Chinese | 81 | 22.3 | – | 22.3 |
| Greek | 51 | 14.0 | 14.0 | |
| Indian | 29 | 8.0 | – | 8.0 |
| Thai | 24 | 6.6 | – | 6.6 |
| Polish | 23 | 6.3 | 6.3 | |
| Moldavian | 13 | 3.6 | – | 3.6 |
| Romanian | 9 | 2.5 | – | 2.5 |
| German | 7 | 1.9 | 1.9 | |
| Egyptian | 5 | 1.4 | – | 1.4 |
| Yugoslav | 3 | 0.8 | – | 0.8 |
| Moroccan | 2 | 0.5 | – | 0.5 |
| Burmese | 1 | 0.3 | – | 0.3 |
| Russian | 1 | 0.3 | 0.3 | |
| Turkish | 1 | 0.3 | – | 0.3 |
| Ukrainian | 1 | 0.3 | – | 0.3 |
| Sri Lankan | 1 | 0.3 | – | 0.3 |
| Total | 364 | 100.0 | 22.5 | 77.5 |

Note
Great efforts have been put into identifying the nationalities of deceased seamen. Many maritime organizations, including LR, ITF, and INTERCARGO, have been contacted. The main sources for nationalities of seamen are obtained from ITF, where crew's lists of ships that have been issued Blue Certificates by ITF are maintained, ITF do not hold all information for ships that have not been issued Blue Certificates. Another difficulty for the exercise is, in some accidents, some crew members were rescued without recording their nationalities. Therefore, the nationalities of the deceased crew members can not be identified either. Only 364 seamen's nationalities were identified compared with a total of 813 deceased seamen from bulk-carrier disasters from 1990–98.[19,20]

### 6.4 Beneficial owners of open registry ships

The open registry regime enables shipowners to register their ships under foreign flags to escape the effective control and administration of their home registries. This regime results, in general, not only in a poor safety record of ships, but also in high fatality and mortality rates, as discussed previously.[25] Most open registry countries are small and developing nations, and registration revenue is usually the main aim of such a regime. Often, they have no facilities and expertise, or sometimes no intention to maintain effective control and administration of these ships. The true nationalities behind these open registries are shown in Table 20.5 with Greece and Japan heading the list.

The oil disaster from the broken *Erika*,[26] an open registry ship owned and classified by a traditional maritime nation, again drew governments' attention on the open registry regime. A key question was again raised on open registry, i.e. who were the users of the open registry regime? In 1997, there were a total of 11,084 ships flying open registry flags, 72 per cent of which were owned by IMO council members, and 80 per cent owned and controlled by developed maritime nations. Open registry is widely considered a weak link in the maritime safety chain.[10] Those beneficial countries should consider what role and responsibility they should take to improve the situation. The open registry scheme can not survive without support from beneficial countries.

### 6.5 The safety of domestic and inland ferries

The study reveals that, in 1996 alone, more than 2,562 people died due to accidents involving domestic or coastal ferries, which without exception were in developing countries, and most of which were less than 500 g.t. Bangladesh alone, from April 1986 to May 1999, saw a total of 1,650 people die in inland waterway accidents (Table 20.4). However, it is noted that the IMO regulations do not apply to ships of less than 500 g.t., ships not propelled by mechanical means, wooden ships, and ships not normally

*Table 20.4* Bangladesh ferry disasters (1986–99)

| Name of ship | Places of accidents | Accident | Dates | Fatalities |
| --- | --- | --- | --- | --- |
| 1 *Atlas Star* | Sitalakhya River | Sink | 20/04/86 | 200 |
| 2 *Samia* | Meghna River | Sink | 25/05/86 | 600 |
| 3 *Haisal* | Dhaleswari River | Sink | 27/12/88 | 200 |
| 4 *XXX* | Unknown | Collision | 28/01/90 | 150 |
| 5 *Dina* | Meghna River | Sink | 20/08/94 | 300 |
| 6 *Dwipkaya* | Lkhsmipur | Sink | 08/05/99 | 200 |
| Total | | | | 1,650 |

Source: Lloyd's list, 11 May 1999.

Table 20.5 True nationality behind open registries (1997)

| States (Areas) | Ships | Members of IMO Council | Developed states (areas) | Developing states |
|---|---|---|---|---|
| 1 Greece | 1,970 | 1,970 | 1,970 | – |
| 2 Japan | 1,862 | 1,862 | 1,862 | – |
| 3 Germany | 926 | 926 | 926 | – |
| 4 US | 703 | 703 | 703 | – |
| 5 Norway | 543 | 543 | 543 | – |
| 6 HK | 517 | – | 517 | – |
| 7 UK | 485 | 485 | 485 | – |
| 8 China | 344 | 344 | – | 344 |
| 9 Korea | 284 | 284 | 284 | – |
| 10 Taiwan | 249 | – | 249 | – |
| 11 Singapore | 224 | 224 | 224 | – |
| 12 Denmark | 198 | – | 198 | – |
| 13 Russia | 194 | 194 | 194 | – |
| 14 Switzerland | 183 | – | 183 | – |
| 15 Italy | 150 | 150 | 150 | – |
| 16 Belgium | 142 | – | – | – |
| 17 Sweden | 142 | 142 | 142 | – |
| 18 France | 99 | 99 | 99 | – |
| 19 Finland | 56 | 56 | 56 | – |
| 20 Saudi Arabia | 54 | – | – | 54 |
| 21 Unknown | 1,759 | – | – | – |
| Total | 11,084 | 7,982 | 8,785 | 398 |
| % | 100 | 72.01 | 79.26 | 3.59 |

Source: UNCTAD 1997[31]

engaged on international voyages (SOLAS, Regulation 3). This leaves a grey area for safety of life at sea. Ships, especially coastal ferries in developing nations, have even not been properly registered and named. However, these ferry disasters, which are not even widely known and discussed, are considered a purely domestic rather than an international matter, and not the concern of the IMO, which apparently needs to be examined and improved.

## 7 Conclusion

There is an urgent need for co-operation on prevention of occupational accidents to seamen between the IMO and the ILO. Inland ferries and passenger ships are considered other weak areas in the coverage of the IMO safety regulations, which have been left entirely to the nations concerned. However, the lack of technology and regulatory machinery in some of the least developed countries leaves a black hole regarding the safety of life on board ship.

# Acknowledgement

Thanks go to the IMO, the ILO, the IMB, the ITF, INTERCARGO, Paris MOU on PSC and LRS for provision of relevant information.

# References and notes

1 Singh, N., 1973, *British shipping laws, Vol. 8, International convention of merchant shipping*, 2nd edn (London: Steven & Sons).
2 Doganis, R.S. and Metaxas, B.N., 1976, *The impact of flags of convenience*. A study sponsored by the Social Science Research Council and carried out at Polytechnic of Central London and Ealing Technical College, London.
3 McConville, J., Glen, D.R. and Dowden, J., 1998, UK seafarer analysis 1997, *International Transport Management*, 1, 3.
4 BIMCO/ISF, 1995, *The worldwide demand for and supply of seafarers* (Warwick: The Institute for Employment Research at University of Warwick).
5 Goss, R.O., Nicholls, C. and Pettit, S.J., 1991, Seamen's accidental deaths and injuries world-wide – a methodology and some estimates, *Journal of Navigation*, 44, 2, 271–5. Li, K.X., 1998, Seamen's accidental deaths worldwide: a new approach. *Maritime Policy & Management*, 25, 149–55.
6 Kitchen, J.S., 1980, *The Employment of Merchant Seamen* (London: Croom Helm) 1.
7 Chapman, P.K., 1992, *Trouble on board: the plight of international seafarers* (Ithaca, New York: IRL Press) 110.
8 In this study, open registries include the Bahamas, Bermuda, Cyprus, Liberia, Malta, Panama, Vanuatu, Antigua and Barbuda, Barbados, Belize, Cambodia, Gibraltar, Honduras, Marshall Islands, Mauritius, Saint Vincent, Sri Lanka and Tuvalu, based on the definition and criteria set out in Li, K.X. and Wonham, J., 1999, New developments in ship registration, *The International Journal of Marine and Coastal Law*, 14, 137–54; and Li, K.X., 1999, The safety and quality of open registers and a new approach for classifying risky ships, *Transportation Research Part E*, 35, 135–43.
9 UNCTAD, 1979, *Beneficial ownership of open-registry fleets: reports by the UNCTAD Secretariat*, TD/222/Supp (Geneva: UNCTAD).
10 Li, K.X. and Wonham, J., 1999, Who is safe and who is at risk: a study of 20-year-record on accident total loss in different flags, *Maritime Policy & Management*, 26, 137–44.
11 Li, K.X. and Wonham, J., 1999, A method for estimating world maritime employment, *Transportation Research Part E*, 35, 183–9.
12 Schilling, R.S.F., 1996, Trawler fishing: an extreme occupation, *American Journal of Industrial Medicine*, 59, 406–10.
13 The comparison would be more convincing if the data on other risk industries would have been collected for the same persons as that from the maritime industry (1962–88). However, the comparison as it stands is still meaningful. The difference between the fatality rates of other industries and that of the maritime industry would have been even larger, bearing in mind accident fatalities have declined during the last three decades as a result of technology and improvements of safety regulations and management.
14 UKCS (UK Chamber of Shipping), 1994, *Digest of accident and health statistics*; Nielsen, D. and Roberts, S., 23, 1, 71–80, 1999, Fatalities among the world's merchant seafarers (1990–1994), *Marine Policy*; Wickramatillake, H.D., 1997, Health at sea, *The Safety & Health Practitioner*, March; Roberts, S., 1997, Mortality among seafarers on the British merchant fleets (1986–95): findings from

an on-going research study. Seafarers International Research Centre, University of Wales, Cardiff, UK. Baines, B., 1981, Mental Health of officer personnel in Indian merchant marine, PhD Dissertation, University of Bombay, Bombay.

15  IMO, 1999, IMO News, No. 3 and 4 (London: IMO).

16  The IMO/ILO have developed a process for investigating human factors, which has partly addressed some of the problems revealed by this chapter, see IMO Resolution A.884(21), Appendix.

17  *Lloyd's List*, 8 November 1999.

18  Thanks go to the International Maritime Bureau (IMB) for provision of relevant data.

19  Thanks go to Mr Greta Powell at INTERCARGO for provision of their publication of Bulk Carrier Casualty Report (1990–1997), Intercargo, 1998, *Bulk carrier casualty report, analysis of total loss and fatality statistics for dry bulk carriers 1970–1997* (London: Intercargo), and Mr John Crilley at Lloyd's Registry.

20  Thanks go to Mr T. Holmer and Ms D.L. He at ITF for the assistance in this regard.

21  Li, K.X. and Wonham, J., 1999, Who mans the world fleet? A follow up to the BIMCO/ISF manpower survey, *Maritime Policy & Management*, 26, 295–303.

22  IMO Resolution A.713(17) 'Safety of Ships Carrying Solid Bulk Cargoes', adopted on 6 November 1991.

23  IMO News, No. 2, 1999, page 3.

24  Notably, all seamen who died with *Derbyshire* were UK citizens.

25  Open registry fleets in general have maintained the highest total lost rate (Li and Wonham,[10] p. 143). However, the Marshall Islands, Liberia, Vanuatu and Barbados had better records than the world average in terms of total loss rates; Bermuda, the Marshall Islands, Liberia, Bahamas were better than the world average detention rate also.

26  *Lloyd's List*, 3 February and 4 April 2000.

27  UK Department of Trade, 1962–88, *Casualties to vessels and accidents to men* (London: HMSO).

28  Transport Committee of House of Commons, 1988, *Decline of the UK registered merchant fleet*. pxv.

29  Department of Transport, *Transport statistics report – Merchant Fleet Statistics 1996* (London: The Stationery Office), 1994, 29.

30  International Maritime Bureau (IMB), 1999.

31  UNCTAD, 1997, *Review of Maritime Transport 1997* (Geneva: UNCTAD).

# 21 Raising world maritime standards

*His Excellency William O'Neil*

The raising of world standards in the maritime industry is a subject that certainly provokes some interesting and perhaps controversial possibilities. IMO's 162 member states, together with more than 60 non-governmental organisations and 30 intergovernmental organisations can generate a myriad of different viewpoints on any given subject. This is one of the organisation's greatest strengths, because this combined capacity of expertise, or brainpower, means that any standard or guideline, code of practice or any other matter dealt with by IMO is exposed to the closest scrutiny before it is supplied to almost 100 per cent of shipping engaged in international trade and that there is no other organisation, either international or regional that has that capability. This is why measures such as the SOLAS Convention, MARPOL, STCW, Collision Regulations and so on, have come to really define the very essence of shipping today. And to a great extent, shipping's technical, operational and administrative profiles are all shaped by the developments which stem from the work of IMO.

There certainly can be no doubt that in matters of safety and environmental protection and security, the shipping industry does look to IMO as the leader in creating and raising standards, and in my mind there is no question that that leadership is being provided. IMO's response to serious accidents had been swift, and it's been decisive. Its current proactive policy has created a regulatory infrastructure that covers everything from measures designed to prevent casualties and accidents, and to minimise the damage to the environment through measures which are aimed at ensuring an effective response when accidents do happen, and onto those activities which have created the compensation regime which ensures that innocent victims of pollution and other accidents can receive what is certainly, in a lot of our minds, considered to be adequate recompense.

Under IMO's leadership standards have been raised to unprecedented levels. The World's fleet now consists of around 88,000 ships, and the vast majority of these operate for their entire lives, safely, cleanly and efficiently within the sound regulatory framework that has been built up over the years by IMO. And I will come back to your point on why the industry requires the regulations in a little while when I finish.

The casualty rate for all types of vessels has plummeted over successive decades. It is worthwhile reminding ourselves that shipping plays a massive part in our collective wellbeing. Millions of tonnes of raw materials, finished products and goods are transported economically, cleanly and without mishap every day of the week, underpinning the global economy and fulfilling international trade. It would, I think, astonish most people to discover that in terms of average annual loss rates per million flight or voyage hours at risk, the loss rate for commercial aircraft is three times that for merchant ships, and that since 1988 the merchant ship loss rate per 1,000 units at risk has been consistently lower than the commercial aircraft rate. Yet, despite all of this, shipping continues to be plagued by a dismal public image. The fact that the vast majority of shipowners fulfil their obligations conscientiously certainly tends to be overshadowed by a few, and I stress 'few' irresponsible operators who skimp on maintenance and training, ignore good practice, while attempting to hide behind a veil of corporate secrecy and obscure transnational chains of ownership and responsibility.

Consider the sequence of events that usually follows a major shipping accident. The first priority, quite rightly, is always to save lives. To rescue the crew and any passengers that may be involved and take them to some place of safety. The next concern is pollution abatement, to protect the environment from the ship's cargo and fuel. Meanwhile, the shipowner stands back from the situation, frequently turns the issue over to a P & I club, and essentially runs for cover, and then the finger pointing begins. Some call for the Master to be thrown into jail, or the First Officer to be charged with negligence. Some blame the owner for skimping on repairs and maintenance. But the owner's reply is usually that 'the ship was in class'. So maybe it's the Classification Society's Inspectors that are to blame. Or shoddy workmanship in construction, or at a repair yard.

Sooner or later these ripples spread still further. The flag state is accused of not demanding sufficiently high standards, the coastal state of offering no place of refuge, and the port state inspectors of missing obvious failings.

Soon afterwards comes the inevitable political outcry. The call goes for tougher regulations and for regional solutions to international problems. It is a good opportunity for politicians to appear strong and decisive and to engage public opinion on their side. The shipping industry meanwhile complains about the knee-jerk responses, and that these are not thought out, that they will bring market disruption and chaos, and anyway will do little to actually improve the situation.

While I certainly would agree that most politicians are not technical experts in shipping, and also that some do not always appear to get the best advice, at the end of the day you have to ask who can blame them for stepping in to regulate the industry if, after decades of trying to regulate itself, the perception is one of disorder and uncaring operators who are insensitive to the safety and environmental concerns of the public at large? And of course, through all this, the media has a field day. Our television screens are filled with the most vivid and emotive images of the distressed ship, of polluted beaches, and of innocent wildlife damaged and destroyed. Then the story spreads to encompass the human element. Livelihoods based on fishing and tourism are threatened. Who is going to compensate these people, and why is there never enough to go around?

The public, not surprisingly, reacts in horror. How can this possibly happen in the twenty-first century, despite the astonishing legacy of scientific and technical advances which have been bequeathed to use? Whenever there is a serious shipping accident there is inevitably a clamour to find someone, or somebody, to place the blame on. But most accidents are the result of a complex chain of events and causes which individually would not be catastrophic, but when they do occur together can have disastrous consequences.

When no obvious scapegoat emerges it is usually the regulatory regime that eventually comes under the spotlight. Of course, any lessons to be learned must be identified and acted upon quickly and decisively. But more regulation or even tighter regulation is frequently not the sole, or even the best solution to the perennial problem of raising standards of safety and improving environmental protection in this shipping industry. For example, all single hull tankers are to be phased out. We know that. It is one of the provisions of the MARPOL Convention. And the only question being currently addressed is whether or not the timetable for doing so should be accelerated once again. No new single hull tankers have been built since the mid-1990s, and so there will come a time, sooner or later, when all oil tankers will have double hulls. Now what will that mean? Will it mean no more oil spills? Will it signify the final solution to the problem of tanker safety? Of course not. It will simply mean that the next generation of tanker to spill its cargo on a beach somewhere because it has been either poorly built, poorly maintained, poorly repaired, or poorly operated, or any combination of these, will be a double hull ship. Because while a double hull may offer some additional protection in certain circumstances, it is just as vulnerable to malpractice as any other.

My point is that all the regulation in the world will not prevent accidents from occurring, and so we must address the need to pursue sound maintenance and operational practices, if casualty rates are to be improved still further. The regulatory regime definitely has its part to play, but we should be clear, the reasons that ships fall apart, flounder, collide with each other, run aground, break up, catch fire, or whatever else may befall them, is rarely, if ever, because there is something fundamentally wrong with the regime. In the vast majority of cases it is because somebody, somewhere along the line, did not take a proper action to avert a problem, or did something wrong, whether through laziness or ignorance, greed, malice, fatigue, negligence, whatever the fact may be, which incidentally stresses how vital it is that IMO's focus on people, as demonstrated by the introduction of the International Safety Management Code, and the updating of the Training Convention, must be sustained.

IMO does not operate ships, sub-standard or otherwise. It may set the benchmarks which has given the term sub-standard some meaning, but it's necessary to look elsewhere if you want to identify who puts sub-standard ships to sea and who allows them to do so. The primary responsibility for the quality of the ships and of the crews who are involved in running them, lies with their owners and their operators. At the moment, there is a sharp focus on the age of ships, and as I have just said, we are currently phasing out older takers or single hull construction and demanding that tankers and bulkers undergo a most robust survey regime when they reach a certain age. I personally think this principle should be applied to all vessels when they reach 15 years, and should be sustained during the balance of the operating lives.

But in the wider context what can we do to push standards still higher? The member states of IMO and particularly the parties to the various conventions, create and adopt standards which they undertake to apply to the ships that fly their flag. But really the process is like a ferry boat journey with several stops along the way. When the journey begins and new ideas are conceived and a new convention is being developed then everyone is happy to crowd on board, but many seem to disembark before the next leg of the journey, that is the ratification of the convention. And still more have left the boat by the time we get around to the chore of full implementing the convention and all of its requirements. And when it comes to making sure ships maintain their quality throughout their working lifetimes our ferry is almost empty.

The responsibility for safety and environmental standards in international shipping is currently handled through a tripartite agreement. Standards are adopted by IMO, implemented by shipowners and operators, and enforced or policed by states, whether as flag states or port states. IMO does not participate in that enforcement activity, it lacks both resources and more importantly the mandate to undertake these tasks. But if dramatic improvements in standards in the shipping industry are to be achieved I believe even further changes will have to be introduced. IMO needs the authority to verify that flag states really do implement conventions fully and properly. The so-called white list of parties to the revised STCW Convention that are deemed to be giving full effect to its provisions is a case in point. It certainly makes a step in the right direction but I believe that we need to go further. Other conventions should have similar 'performance clauses'. There should be a provision for sanctions and penalties which may be applied if convention requirements are not adhered to. IMO needs an audit system whereby its members' performance can be properly monitored and such a scheme is currently being assessed by the organisation. A requirement should be introduced that all accident investigation reports must be submitted to IMO with all possible speed, and not just at the discretion of the flag state concerned.

The job of raising world standards in the shipping industry will never be completed, never be finished because we do live in a very complex and ever-changing world and therefore there is still a great deal which needs to be attended to, both in the short and in the longer term. But we should not let that obscure the huge advances that in fact have already been made. I think that we are in the right lane and that we have travelled some considerable distance down it. The emphasis will be on continuing to focus on the human element, on continuing to stress the importance of a culture of safety and environmental concern that will permeate the whole shipping industry from top to bottom.

Ultimately the real key to further wholesale improvements in shipping standards lies in changing some of the fundamental philosophical traditional and structural premises on which the industry has been built. If, in the future, shipowners have to compete on quality as much as they currently do on cost, then standards will inevitably rise. But to see this happen there will have to be an acceptance of the fact that a safe, secure maritime transport system does in fact provide a genuine value-added link in the manufacturing and supply process. And while added value does come with a price tag attached to it, those costs would be greatly outweighed by the benefits to the shipping industry and to society as a whole.

# Index

9 780415 649254